Community-Based Adaptation

Community-based adaptation (CBA) to climate change is based on local priorities, needs, knowledge and capacities. Early CBA initiatives were generally implemented by non-government organisations (NGOs), and operated primarily at the local level. Many used 'bottom-up' participatory processes to identify the climate change problem and appropriate responses.

Small localised stand-alone initiatives are insufficient to address the scale of challenges climate change will bring, however. The causes of vulnerability – such as market or service access, or poor governance – also often operate beyond the project level. Larger organisations and national governments have therefore started to implement broader CBA programmes, which provide opportunities to scale up responses and integrate CBA into higher levels of policy and planning.

This book shows that it is possible for CBA to remain centred on local priorities, but not necessarily limited to work implemented at the local level. Some chapters address the issue of mainstreaming CBA into government policy and planning processes or into city or sectoral level plans (e.g. on agriculture). Others look at how gender and children's issues should be mainstreamed into adaptation planning itself, and others describe how tools can be applied and finance delivered for effective mainstreaming.

This book was published as a special issue of *Climate and Development*.

Hannah Reid is a Research Associate at the International Institute for Environment and Development. She has over 13 years of experience working on climate change and development, with particular focus on how to help those who are most vulnerable to climate change cope with its impacts, and on ecosystem-based approaches to adaptation.

Community-Based Adaptation

Mainstreaming into national and local planning

Edited by
Hannah Reid

Routledge
Taylor & Francis Group

LONDON AND NEW YORK

First published 2016
by Routledge
2 Park Square, Milton Park, Abingdon, Oxon, OX14 4RN, UK

and by Routledge
711 Third Avenue, New York, NY 10017, USA

First issued in paperback 2017

Routledge is an imprint of the Taylor & Francis Group, an informa business

British Library Cataloguing in Publication Data
A catalogue record for this book is available from the British Library

ISBN 13: 978-1-138-29493-6 (pbk)
ISBN 13: 978-1-138-96342-9 (hbk)

Typeset in Times New Roman
by diacriTech, Chennai

Publisher's Note
The publisher accepts responsibility for any inconsistencies that may have arisen
during the conversion of this book from journal articles to book chapters, namely
the possible inclusion of journal terminology.
The publisher acknowledges that the original special issue was available open
access.

Disclaimer
Every effort has been made to contact copyright holders for their permission to
reprint material in this book. The publishers would be grateful to hear from any
copyright holder who is not here acknowledged and will undertake to rectify any
errors or omissions in future editions of this book.

Contents

CONTENTS

Citation Information

The chapters in this book were originally published in *Climate and Development*, volume 6, issue 4 (October 2014). When citing this material, please use the original page numbering for each article, as follows:

Chapter 8
A review of decision-support models for adaptation to climate change in the context of development
John Jacob Nay, Mark Abkowitz, Eric Chu, Daniel Gallagher and Helena Wright
Climate and Development, volume 6, issue 4 (October 2014) pp. 357–367

Chapter 9
Enablers for delivering community-based adaptation at scale
Elizabeth Gogoi, Mairi Dupar, Lindsey Jones, Claudia Martinez and Lisa McNamara
Climate and Development, volume 6, issue 4 (October 2014) pp. 368–371

Chapter 10
Mainstreaming children's vulnerabilities and capacities into community-based adaptation to enhance impact
Paul Mitchell and Caroline Borchard
Climate and Development, volume 6, issue 4 (October 2014) pp. 372–381

Chapter 11
Knowledge flows in climate change adaptation: exploring friction between scales
Clare Stott and Saleemul Huq
Climate and Development, volume 6, issue 4 (October 2014) pp. 382–387

Chapter 12
Up-scaling finance for community-based adaptation
Adrian Fenton, Daniel Gallagher, Helena Wright, Saleemul Huq and Charles Nyandiga
Climate and Development, volume 6, issue 4 (October 2014) pp. 388–397

For any permissions-related enquiries please visit
http://www.tandfonline.com/page/help/permissions

Notes on Contributors

Mark Abkowitz is Professor of Civil and Environmental Engineering at Vanderbilt University, Nashville, Tennessee, USA. He specialises in enterprise risk management, hazardous materials transportation safety and security, assessing the impacts of energy choices and climate change and the strategic and operational deployment of intelligent transportation systems.

Florencia Almansi is based at the Instituto Internacional de Medio Ambiente y Desarrollo-América Latina, in Buenos Aires, Argentina.

Edidah Ampaire is the Project Coordinator for the CCAFS Policy Action for Climate Change Adaptation Project, based at the International Institute of Tropical Agriculture, Kampala, Uganda. She also supports other projects in integrating gender and participatory research methods.

Diane Archer is a Researcher in the Human Settlements Group at the International Institute for Environment and Development. Her current work is on researching Asian cities and climate change and on humanitarian responses to crises in urban areas.

Jessica Ayers is a Senior Adviser in the Department of Energy and Climate Change for the UK Government, based in London, UK. She holds a PhD in International Development and Environmental Change from the London School of Economics, UK.

Caroline Borchard is Regional Climate Change Coordinator for Plan International, Bangkok, Thailand.

Eric Chu is a Doctoral Researcher in Environmental Policy and Planning in the Department of Urban Studies and Planning at MIT, Cambridge, Massachusetts, USA. His research looks at the multilevel politics of climate change adaptation and development planning in Indian cities.

Cristina Coirolo is a Survey Co-ordinator and Social Scientist at Columbia University, New York, USA. She is working on a project titled 'Building Resilience to Storm Surges and Sea Level Rise: A Comparative Study of Coastal Zones in New York City and Boston'.

Michael DiGregorio is The Asia Foundation's Country Representative in Vietnam. He has extensive professional and academic experience focused on Vietnam. He has worked with a variety of governmental, academic, business and international non-government organisations in the country over the past 22 years.

Mairi Dupar is Global Public Affairs Co-ordinator for the Climate and Development Knowledge Network (CDKN). Her work involves commissioning articles, films and events for CDKN and coordinating a team of knowledge management staff across CDKN's global hub in London, UK, and regional offices in Asia, Africa and Latin America.

Arif M. Faisal is an Environmental Specialist at the Asian Development Bank, based in Bangladesh. He has 15 years of professional experience in the areas of climate change, environment and development.

Adrian Fenton is a Research Student based in the Centre for Climate Change Economics and Policy, University of Leeds, UK. His project title is 'Climate change adaptation and microfinance: a potential win-win scenario?'

Daniel Gallagher is Junior Professorial Associate for the Adaptation Fund Board Secretariat, Washington DC, USA. He has 3 years of professional experience as an engineering consultant in technical design reviews of energy and infrastructure projects in the UK and as technical lead on community-based infrastructure projects in Ecuador and El Salvador.

Elizabeth Gogoi is a Project Manager with the Climate and Development Knowledge Network, based in New Delhi, India. She has a background in climate change and in development policy-making and research. This includes working in the European Parliament on the negotiations of the 2009 EU Climate Change Package and as a researcher in the Overseas Development Institute (ODI) looking at EU development cooperation.

Saleemul Huq is a Senior Fellow at the International Institute for Environment and Development. His expertise lies in the links between climate change and sustainable development, particularly in developing countries. He is involved in building negotiating capacity and supporting the engagement of the Least Developed Countries (LDCs) in the UN Framework Convention on Climate Change.

Syed Tanveer Hussain is CEO of the Climate Change Company, based in Dhaka, Bangladesh. He is involved in adaptation, mitigation, the buying and selling of carbon credits, low carbon emission strategies as well as issues relating to renewable energy and energy efficiency. He has previously been Consultant to the World Bank and Secretary to the Bangladeshi government.

M.L. Jat is a Senior Scientist (Cropping Systems Agronomist) in the Global Conservation Agriculture Program, based at the International Maize and Wheat Improvement Centre in New Delhi, India. He is currently responsible for developing CA-based management solutions to address issues of resource degradation, abiotic stress and climate change in South Asia.

Lindsey Jones is a Research Fellow for the Overseas Development Institute, based in London, UK. His principle areas of research interest include promoting effective principles of adaptation, disaster risk reduction and resilience; understanding factors that contribute to adaptive capacity; strengthening national/local adaptation strategies and institutional capacity building; and assessing the impacts of climate change on access, entitlements and management of natural resources.

Shusmita Khan is based at Eminence, the Center for Health and Development Intelligence, in Dhaka, Bangladesh. Her skills and expertise lie in public health, intervention studies and health care.

Gernot Laganda is a Climate Change Adaptation Specialist for the International Fund for Agricultural Development, based in Rome, Italy.

Claudia Martinez is Executive Director of E3-Ecology, Economics and Ethics, Representative for the Colombia Climate and Development Alliance (CDKN), Manager and Co-ordinator of a regional climate adaptation program focused on the Huila Department (Huila 2050) and Co-founder of 'Sistema B', a regional initiative to promote businesses that use the power of the market to create social and environmental benefits.

Lisa McNamara is Knowledge Management and Partnership Coordinator, Africa, for the Climate and Development Knowledge Network. She holds an MSc in Environmental Science with distinction; her thesis focuses on climate change adaptation governance in cities.

Paul Mitchell is a Climate Change and Sustainable Development Consultant for adapt-develop, based in Australia. He provides consulting services to government and non-government organisations on climate change and sustainable development issues.

John Jacob Nay is a PhD student at Vanderbilt University, Nashville, Tennessee, USA, and a Fellow at the Vanderbilt Institute for Energy and Environment. He has also been awarded various honors, including a Fulbright Scholarship and the Senator John Warner Public Leadership Research Award.

Charles Nyandiga is a Programme Advisor at the United Nations Development Programme, based in New York City, USA.

Max Olupot is based at the African Forum for Agricultural Advisory Services, based in Kampala, Uganda.

Atiq Rahman is Executive Director of the Bangladesh Centre for Advanced Studies, Dhaka, Bangladesh. He is a prominent environmentalist, scientist, development expert and a visionary thinker in South Asia. He is well-known worldwide for his pioneering role and contribution to environment and nature conservation, climate change, poverty alleviation and sustainable development.

NOTES ON CONTRIBUTORS

Bimal Raj Regmi is a PhD candidate in the School of Social and Policy Studies at Flinders University, Adelaide, Australia. He is interested in climate change and community-based adaptation, participatory research and development, climate change and natural resources.

Hannah Reid is a Research Associate at the International Institute for Environment and Development. She has over 13 years of experience working on climate change and development, with particular focus on how to help those who are most vulnerable to climate change cope with its impacts, and on ecosystem-based approaches to adaptation.

Debra Roberts founded and heads the Environmental Planning and Climate Protection Department of eThekwini Municipality, Durban, South Africa. Her key responsibilities in this post include overseeing the planning and protection of the city's biodiversity and natural resource base; directing and developing the municipality's Climate Protection Programme; and ensuring that biodiversity and climate change considerations influence all aspects of planning and development in the city.

Dalia Shabib is the Community Engagement Co-ordinator for Terre des Hommes Italia, based in Lebanon.

Divya Sharma is a Fellow and Area Convener at the Centre for Research in Sustainable Urban Development and Transport Systems, at the Sustainable Habitat Division at TERI University, New Delhi, India. She is a trained architect and an urban planner, and holds a PhD in climate change adaptation in cities.

Cassandra Star is Senior Lecturer in Public Policy at Flinders University, Adelaide, Australia. Her research centres on environmental politics and policy, with a focus on the politics of climate change and on the role, actions and influence of non-government organisations in this arena. She is particularly interested in both the political influence of the movement as well as the formal and informal networks and social learning that occur between groups in the non-government sector around climate change issues.

Claire Stott is a Visiting Researcher at the International Centre for Climate Change and Development in Dhaka, Bangladesh. She holds an MSc from University College, London, UK.

Denia Syam is National Network Coordinator for the Asian Cities Climate Change Resilience Network, based in Jakarta, Indonesia.

Sonja Vermeulen is Head of the Research Program on Climate Change, Agriculture and Food Security, based at the University of Copenhagen, Denmark. Her work has spanned the natural and social sciences across the fields of forestry, agriculture and natural resource management. Her career has bridged academic and applied research, with a strong focus on linking science with public policy and private sector strategies, particularly in Africa and Asia.

Helena Wright is a Research Postgraduate in the Centre for Environmental Policy at Imperial College London, UK. She has contributed to various publications, articles and books, writing on climate change adaptation, mitigation, food security and technology transfer. Her research interests include climate finance effectiveness, international negotiations on climate change and climate change impacts in developing countries.

INTRODUCTION

Mainstreaming community-based adaptation into national and local planning

Hannah Reid[a] and Saleemul Huq[a,b]

[a]International Institute for Environment and Development, London, UK; [b]ICCCAD, Dhaka, Bangladesh

Community-based adaptation (CBA) to climate change can be defined as 'a community-led process, based on communities' priorities, needs, knowledge, and capacities, which should empower people to plan for and cope with the impacts of climate change' (Reid, Cannon, Berger, Alam, & Milligan, 2009). Early CBA initiatives were generally implemented by non-government organizations, and operated primarily at the local level. Emphasis was placed on applying 'bottom-up' participatory processes to identify the climate change problem and appropriate local responses to this problem (Ayers & Forsyth, 2009).

As realization grew about the scale of the problems humanity will face as a result of climate change, it became clear that small localized stand-alone initiatives were not enough to respond to the challenges (Reid, 2014; Schipper, Ayers, Reid, Huq, & Rahman, 2014). Dodman and Mitlin (2013) argued that while CBA was strong on emphasizing participatory processes, insufficient attention was given to building up links with political structures above the level of the settlement. Others stressed that many climate change impacts could not be managed through local adaptation, and that the multiple causes of vulnerability included market or service access or good governance beyond the project level (Dixit, McGray, Gonzales, & Desmond, 2012; Ensor & Berger, 2009).

Increasingly, larger multilateral and bilateral agencies, national governments, and representatives from major government and non-government agencies were taking an interest in CBA and starting to implement larger scale programmes. Early international CBA conferences run by the International Institute for Environment and Development, the Bangladesh Centre for Advanced Studies, and local partner organizations, were dominated by non-government organizations that pioneered much of the early CBA work. The theme of the fifth international CBA conference held in 2011, however, was 'Scaling Up: Beyond Pilots' reflecting a growing interest in scaling up CBA activities. The range of stakeholders attending was also much broader than at earlier conferences (Haider & Rabbani, 2011).

Whilst non-government organizations have done much to promote learning on CBA and implement activities at the grassroots level, stronger engagement with a wider group of stakeholders, particularly governments, provides opportunities to move away from isolated pilot projects and integrate CBA into levels of policy and planning to an extent that non-government organizations could not do (Huq & Ayers, 2008; Pelling, 2011; UNDP/UNEP, 2011). Klein, Schipper, and Dessai (2005) argue that mainstreaming adaptation into local, regional and national government structures and processes in this way is more sustainable, effective and efficient than designing and managing policies separately from ongoing activities. It may also protect adaptation activities from stakeholders who see them as a threat or do not support their aims, and help avoid conflict with existing policies (Lebel et al., 2012).

Experience from a number of programmes such as the Global Environment Facility's Small Grants Programme executed by the United Nations Development Programme was providing evidence that CBA initiatives could operate at scale, for example, through mainstreaming into broader government and non-government policy and planning processes. Operating at scale can lead to tensions and challenges, for example, government structures are notoriously slow to take action and respond to local needs and many have a very chequered history of responding to the needs of the poorest and most vulnerable. But experience has shown that it is possible for CBA to remain centred on the priorities and processes chosen by the community but not necessarily limited to work implemented at the level of the community (Reid & Schipper, 2014).

Arguably the best practical example of mainstreaming CBA into broader planning processes is provided by Nepal. In 2011, the Nepalese government adopted Local Adaptation Plans of Action as the official framework for national adaptation planning (Government of Nepal, 2011). The Government of Nepal had realized that most climate change impacts were felt at the local level, and that there was a disconnect between local and national level planning (namely the National Adaptation Programme of Action) on how best to respond to climate change. The country's long history of community forestry provided a precedent on which to base the work that followed, and policies such as the Decentralisation Act of 1982 provided a supportive legislative framework in which to cluster bottom-up natural resource management and development activities and hence mainstream local adaptation actions into national level planning.

The seventh international CBA conference in 2013 explored the bottlenecks and challenges associated with the theme of systematically 'Mainstreaming CBA into National and Local Planning'. For example, participants identified a need for better integration of CBA with disaster preparedness activities, including early warnings and disaster risk reduction activities. A significant cohort of government representatives attended and shared experiences of mainstreaming CBA into government programmes from The Gambia, Kenya, Bangladesh and Cambodia (Reid et al., 2013). It became clear that whilst core ministries of planning and finance are increasingly becoming involved, countries are finding their own ways of developing strategies to address the impacts of climate change on national development. Trajectories have different starting points and pathways.

The papers published in this special issue have all emerged from presentations made and ideas shared at this seventh international CBA conference. Some address the issue of mainstreaming CBA into government policy and planning processes, for example, at national levels in Bangladesh and Nepal, or at the level of the city or a specific sector such as agriculture. Others look at how gender and children's issues should be better mainstreamed into adaptation planning, including CBA. And others provide examples of how tools can be applied, and finance delivered for effective mainstreaming.

Many CBA practitioners are based in non-government organizations, and increasingly government agencies, where it is difficult to dedicate much time to publishing work in academic journals. This means that much of their knowledge and experience is not shared as widely as it could be. This special issue has channelled support provided by the UK Government's Department for International Development to help some of these practitioners develop papers of a high enough standard to merit publication, and hence share their knowledge more widely.

References

Ayers, J., & Forsyth, T. (2009). Community-based adaptation to climate change: Strengthening resilience through development. *Environment, 51*(4), 22–31.

Dixit, A., McGray, H., Gonzales, J., & Desmond, M. (2012). *Ready or not: Assessing national institutional capacity for climate change adaptation.* Washington, DC: World Resources Institute.

Dodman, D., & Mitlin, D. (2013). Challenges for community-based adaptation: Discovering the potential for transformation. *Journal for International Development, 25*(5), 640–659.

Ensor, J., & Berger, R. (2009). *Understanding climate change adaptation: Lessons from community-based approaches.* Rugby: Practical Action.

Government of Nepal. (2011). *National framework on local adaptation plans for action.* Singha Durbar: Government of Nepal, Ministry of Science Technology and Environment.

Haider, S.S., & Rabbani, G. (2011, March 24–31). Conference proceedings: 5th international conference on community based adaptation (CBA), Bangladesh Centre for Advanced Studies, Dhaka.

Huq, S., & Ayers, J. (2008). *Taking steps: Mainstreaming national adaptation.* London: IIED Policy Brief.

Klein, R.J.T., Schipper, L., & Dessai, S. (2005). Integrating mitigation and adaptation into climate and development policy: Three research questions. *Environmental Science & Policy, 8*, 579–588.

Lebel, L., Li, L., Krittasudthacheewa, C., Juntopas, M., Vijitpan, T., Uchiyama, T., & Krawanchid, D. (2012). *Mainstreaming climate change adaptation into development planning.* Bangkok: Adaptation Knowledge Platform and Stockholm Environment Institute.

Pelling, M. (2011). *Adaptation to climate change: From resilience to transformation.* London: Routledge.

Reid, H. (2014). *Ecosystem- and community-based adaptation: learning from natural resource management.* IIED Briefing. London: IIED.

Reid, H., Cannon, T., Berger, R., Alam, M., & Milligan, A. (Eds.). (2009). *Community-based adaptation to climate change. Participatory Learning and Action 60.* London: IIED.

Reid, H., Coirolo, C., Christensen, K., Fenton, A., Roberts, E., Stott, C., … Wright, H. (Eds.). (2013, April 18–25). *Community based adaptation: Mainstreaming CBA into national and local planning.* Conference proceedings: 7th international conference, Dhaka, Bangladesh. London: IIED/BCAS.

Reid, H., & Schipper, E.L.F. (2014). Upscaling community-based adaptation: An introduction to the edited volume. In E.L.F. Schipper, J. Ayers, H. Reid, S. Huq, & A. Rahman (Eds.), *Community based adaptation to climate change: Scaling it up*, pp. 3–21. London: Routledge.

Schipper, E.L.F., Ayers, J., Reid, H., Huq, S., & Rahman, A. (Eds.) (2014). *Community based adaptation to climate change: Scaling it up.* London: Routledge.

UNDP/UNEP. (2011). *Mainstreaming adaptation to climate change in development planning. A guidance for practitioners.* Nairobi, Kenya: UNDP-UNEP Poverty Environment Initiative.

REVIEW ARTICLE

Mainstreaming climate change adaptation into development in Bangladesh

Jessica Ayers[a], Saleemul Huq[b,c], Helena Wright[d], Arif M. Faisal[e] and Syed Tanveer Hussain[f]

[a]Department of Energy and Climate Change, 3 Whitehall Place, London, UK; [b]International Institute for Environment and Development, London, UK; [c]International Centre for Climate Change and Development, Dhaka, Bangladesh; [d]Centre for Environmental Policy, Imperial College London, UK; [e]Asian Development Bank, Bangladesh Resident Mission, Bangladesh; [f]The Climate Change Company, Dhaka, Bangladesh

The close linkages between climate change adaptation and development have led to calls for addressing the two issues in an integrated way. 'Mainstreaming' climate information, policies and measures into ongoing development planning and decision-making has been proposed as one solution, making a more sustainable, effective and efficient use of resources than designing and managing climate policies separately from ongoing activities. But what does mainstreaming look like in practice? This paper reviews the process of mainstreaming in Bangladesh, one of the countries that has made significant progress on adaptation planning and mainstreaming. The paper begins by making the case for mainstreaming, by exploring linkages and trade-offs between adaptation and development and reviewing the literature on mainstreaming. Second, it considers how to implement mainstreaming in practice, reviewing an existing four-step framework. Examining this framework against the plethora of mainstreaming experiences in Bangladesh, the paper considers how the framework can be used as a tool to review progress on mainstreaming in Bangladesh. The paper concludes that while the framework is useful for considering *some* of the preconditions necessary for mainstreaming, experiences in Bangladesh reflect a much more complex patchwork of processes and stakeholders that need to be taken into consideration in further research.

Note: This paper is adapted with kind permission of the journals from Ayers, Huq, Faisal and Hussain, 2013: Mainstreaming climate change adaptation into development: a case study of Bangladesh. Climate Wires 5(1) pp.37–51.

1. Introduction

Adaptation to climate change has been defined as adjustment in natural or human systems in response to actual or expected climatic stimuli or their effects, which moderates harm or exploits beneficial opportunities (IPCC, 2007). Although the whole world is affected by the impacts of climate change, it is widely accepted those most in need of support for adaptation are the poorest people in developing countries (Adger, Huq, Brown, Conway, & Hulme, 2009; Ayers & Dodman, 2010; Burton, 2004; Huq & Ayers, 2007; IPCC, 2007; Schipper, 2007). This is because although exposure to impacts is driven by climatic hazards, the capacity to adapt to these hazards is determined by factors related to (under) development such as poverty, social and political marginalization, meaning people are unable to cope with both climate and other stresses. For example, individuals and households that have reliable access to food and adequate food reserves, clean water, health care and education will inevitably be better prepared to deal with a variety of shocks and stresses – including those arising because of climate change

(Dodman, Ayers, & Huq, 2009). The links between development and adaptation have resulted in calls to tackle the two issues in an integrated way – to 'mainstream' climate change adaptation into development support and development planning (Huq et al., 2004; Klein, Schipper, & Dessai, 2003, Klein, 2010; Olhoff & Schaer, 2010).

Mainstreaming involves the integration of information, policies and measures to address climate change into ongoing development planning and decision-making (Klein et al., 2003). It is seen as making more sustainable, effective and efficient use of resources than designing and managing policies separately from ongoing activities (Ayers & Huq, 2009a; Klein et al., 2003). In theory, mainstreaming should create 'no regrets' opportunities for achieving development that is resilient to current and future climate impacts for the most vulnerable, and avoid potential tradeoffs between adaptation and development strategies that could result in maladaptation in the future (Ayers & Huq, 2009a; Klein et al., 2003).

But what does mainstreaming look like in practice? As the mainstreaming agenda is taken up by international organizations, developed country agencies and developing country planners, various approaches are emerging. This paper reviews the process of mainstreaming in Bangladesh, one of the countries making significant progress on adaptation planning and mainstreaming. It begins by making the case for mainstreaming, exploring the linkages and trade-offs between adaptation and development and describing the various approaches to mainstreaming from the literature. Second, it considers how to implement mainstreaming in practice, drawing on a conceptual framework originally proposed in Huq and Ayers (2008). Finally, it examines this framework against the plethora of mainstreaming experiences emerging in Bangladesh and considers what can be learnt for informing future research on adaptation mainstreaming.

2. Methodology

This paper is intended as a review rather than as a research paper. The primary purpose of this paper is to review a framework for mainstreaming adaptation originally proposed in Huq and Ayers in 2008. At the time the framework was written, mainstreaming was in its infancy, with little in the way of documented practice. At the time, while other guidelines existed for mainstreaming adaptation into development (described in the next section of this paper), the framework was an early attempt to map what was going on in practice. This paper returns to the framework, and considers the value of the framework against what has been learnt from mainstreaming in practice. We review domestic progress on mainstreaming in Bangladesh, which has taken strides at national and sub-national levels in terms of mainstreaming adaptation. The majority of the paper is based on analysis of existing literature and critiques of adaptation planning and mainstreaming in Bangladesh since the publication of the review, and supports this analysis with interview data conducted with Government, non-governmental organization (NGO) and donor officials engaged in the processes described. Firstly, we describe the way mainstreaming has emerged in Bangladesh, and the numerous approaches that have been taken. Second, we review this experience against the framework, and consider how the framework could be updated and revised in the light of these experiences.

3. Mainstreaming adaptation and development: a framework for analysis

3.1. *Why mainstream? The linkages between adaptation and development*

Historically, climate change adaptation and development have been managed in different arenas. Climate change adaptation emerged as a response to climate change impacts as governed under the United Nations Framework Convention on Climate Change (UNFCCC). The 'ultimate objective' of the UNFCCC is the mitigation of greenhouse gas emissions to prevent 'dangerous' climate change. Thus, adaptation emerged under global governance structures from discussions of climate change impacts and how they could be managed. This has developed into an 'impacts-based' approach to adaptation (Burton, Huq, Lim, Pilifosova, & Schipper, 2002; Ford, 2008), which has resulted in what Klein defines as 'technology-based' interventions such as dams, early-warning systems, seeds and irrigation schemes based on specific knowledge of future climate conditions (Klein, 2008).

However, scholars and practitioners from development and disaster risk reduction fields have repeatedly pointed out that such 'stand-alone' approaches to adaptation targeting very specific climate risks, are unlikely to be effective where they do not also address the underlying factors related to development that make people vulnerable (Adger & Kelly, 1999; Cannon, 2000). During the 1980s, observers from these fields began to draw attention to the link between the risks people face, and the reasons behind their vulnerability to these risks in the first place (Sen, 1999). Such arguments suggested that using the impacts of hazards as the starting point for adaptation to environmental hazards was misguided, because it ignores the ways in which local and wider contexts determine people's *vulnerability* (Blaikie, Cannon, Davis, & Wisner, 1994; Smit & Wandel, 2006).

The development community applied this thinking to climate change adaptation as early as 1987, when the Brundtland Report cited climate change as a major environmental challenge facing development (Ayers & Dodman, 2010). Researchers began to apply theories of social vulnerability to adaptation (Adger & Kelly, 1999), while development agencies began to recognize climate change as a threat to development efforts and poverty reduction (Sperling, 2003). Central to the proposals being put forward was that poverty underpins vulnerability, and therefore good development must be the starting point for adaptation. Burton (2004) suggests that analysing vulnerable communities would reveal an existing 'adaptation deficit', which is the existing capacity of many vulnerable countries and groups to cope with and adapt to *existing* climate risks. Adaptation would need to reduce this deficit to increase people's resilience to climatic variation more generally, before they can adapt to future changes (Burton, 2004). Such insights have led some scholars to conclude that much adaptation simply represents a practical means of achieving sustainable development (Huq & Ayers, 2008).

This has given rise to recommendations to support sustainable livelihoods, improve governance and make institutions more accountable and participatory as part of adaptation support (Klein, 2008; Sperling, 2003). For example, in Vietnam, Kelly and Adger (2009) propose that possible adaptive outcomes from a climate-vulnerability analysis might include: prioritizing poverty reduction; income diversification; and addressing land and common property management rights. Such interventions could well be part of a development programme irrespective of climatic risks. Levina (2007) highlights the potential for the Millennium Development Goals to reduce vulnerability: reducing poverty, providing general education and health services, and providing access to financial markets and technologies will all improve the livelihoods of vulnerable people, and increase their adaptive capacity. An analysis of official development assistance (ODA) activities demonstrated that more than 60% of all ODA could be relevant to facilitating adaptation (Levina, 2007).

The relationship between adaptation and development also works in the other direction: climate change poses a direct threat to the sustainability of development investments. The World Bank estimates that up to 40% of development financed by overseas assistance and concessional loans is sensitive to climatic risk (Burton, Diringer, & Smith, 2006). This not only challenges poverty reduction strategies over the medium term, but also consequently undermines the capacity of the poorest people to adapt (Anderson, 2011). Thus, under climate change, the 'adaptation deficit' will be exacerbated.

Finally, failing to take adaptation into account in development practice can result in maladaptation, where actions or investments create further risks for adaptation (Barnett & O'Neill, 2010). Barnett and O'Neill (2010) describe five key dimensions of maladaptation, including actions that increase greenhouse gas emissions; disproportionately burden the most vulnerable; have high opportunity costs; reduce long-term incentives to adapt; and create path-dependency.

Thus, at least in principle, development and adaptation are now recognized as co-dependent (IPCC, 2007; Olhoff & Schaer, 2010). As adaptation gained prominence under the UNFCCC, its context has shifted from being tied into discussions over impacts and thresholds (Burton, 2004) towards explicit recognition of the role of development in managing adaptation in the scientific and policy guidance emerging from the Intergovernmental Panel on Climate Change (IPCC) and UNFCCC (Schipper, 2006). In the development context, donor agencies are increasingly seeking to 'climate-proof' their investments and make them relevant to the building of adaptive capacity (Tanner, 2008). A review undertaken of interventions labelled as 'adaptation' found that in practice, adaptation and development are not implemented as discreet interventions, but instead lie along a continuum between those that overlap almost completely with development, and those focused specifically on climate impacts (McGray, Hammill, & Bradley, 2007). Accordingly, there is broad agreement within both the climate and development community that an integrated approach to doing adaptation and development makes sense (Ayers & Dodman, 2010; Gupta, 2009; IPCC, 2007).

3.2. *Addressing adaptation and development through mainstreaming*

Integrating adaptation into development is often referred to as 'mainstreaming'. In general terms, mainstreaming refers to integrating an issue into existing (usually development) institutions and decision-making. The term is perhaps best known in relation to 'gender mainstreaming' (Booth & Bennett, 2002). More recently, 'environmental mainstreaming' entered the development policy agenda. This is defined by Dalal-Clayton and Bass (2009) as the informed inclusion of relevant environmental concerns into institutional decisions that drive national and sectoral development policy, rules, plans, investment and action.

Mainstreaming, or ensuring integrated policy-making, therefore has a long history in both development and environmental policy (see Ross & Dovers, 2008). Applied to climate change adaptation, mainstreaming has been proposed as a key avenue through which to address adaptation and development together (Huq et al., 2004; Klein, 2008; OECD, 2009). But mainstreaming in practice can mean different things to different people (Dalal-Clayton & Bass, 2009; Klein, 2008).

First, what are we mainstreaming? Perspectives on this question differ depending on whether we take a technology-based (impacts-based) or a development-based view of adaptation (Klein, 2010). In the technology-based view, mainstreaming largely refers to ensuring that projections of climate impacts are considered in decision-making about investments, so technologies (e.g. drainage systems or crop varieties) are chosen or improved to withstand the future climate. This type of mainstreaming has also been referred to as 'climate-proofing' or 'mainstreaming minimum' (Klein, 2008), and can involve screening of development portfolios through a climate-change lens (Klein et al., 2007). A 'climate-proofing' only approach to mainstreaming has been widely criticized for failing to fully address the underlying drivers of vulnerability; not addressing maladaptation; and not realizing the potential of development interventions to achieve climate resilience (Ayers, Kaur, & Anderson, 2011; Klein, 2008; Seballos & Kreft, 2011). For example, strengthening an embankment to ensure that it can withstand anticipated increases in storm surges will not protect those who cannot afford to reside behind it, and may inadvertently encourage investment and settlement in a climate-vulnerable area.

On the other hand, a vulnerability or development-based view of adaptation gives rise to a more holistic approach, in which in addition to climate-proofing, development efforts deliberately aim to reduce vulnerability by including priorities essential for adaptation. Klein (2010) provides the example of securing water rights for groups exposed to water scarcity during a drought. This latter option takes adaptation responses not as stand-alone or discrete options, but as support to a range of processes that address the underlying drivers of vulnerability: 'Mainstreaming-plus' (Klein, 2010) or 'adaptation as development' (Ayers & Dodman, 2010). It recognizes that adaptation involves many actors, requires an enabling environment with existing financial, legal, institutional, and knowledge barriers to adaptation removed, and involves strengthening capacity of people and organizations to adapt (Klein, 2010). Similarly, Gupta and Van Der Grijp (2010) define climate change mainstreaming as the process by which development policies, programmes and projects are (re)designed and (re)organized, and evaluated from the perspective of climate change mitigation and adaptation. This arguably means assessing how they impact the vulnerability of people (Gupta & Van Der Grijp, 2010).

Second, what – or whose – development are we mainstreaming into? Given that much of the support for adaptation is channelled through the international development institutions, mainstreaming has been discussed from the perspective of development cooperation (OECD, 2009), which means making selected investments of donor agencies climate-proof and also relevant to building adaptive capacity. For example, in response to a call from the G8 in Gleneagles (2005) to climate-proof development assistance, the UK Department for International Development (DFID) piloted climate risk assessments of its development portfolios in Bangladesh, India and Kenya, and selected non-DFID funded water sector programmes in China (Tanner, 2008). The Asian Development Bank (ADB) is climate-proofing agriculture, water resources, infrastructure and transport sector projects in Asia and the Pacific, incorporating adaptation and mitigation components in relevant development projects and providing technical assistance for climate-resilient development. International and national NGOs have also played a key role in mainstreaming, including providing information and pilot projects which feed lessons into broader government processes.

But, only focusing on mainstreaming adaptation into external development assistance does not necessarily take into account the corresponding changes required in the wider national and local institutional environments to ensure that investments are sustainable. As noted by Lebel, Li, and Krittasudthacheewa (2012), the national level provides the overall framework within which sectoral and other sub-national levels operate, and where policy goals from long-term strategies are translated into action plans and budgets.

These two perspectives are not necessarily divergent. Indeed, enabling mainstreaming at the national level should be the ultimate purpose of external assistance. The Paris Declaration on Aid Effectiveness commits all donor agencies to supporting national ownership over the development and implementation of development strategies (OECD, 2005). Yet, observers have cautioned that, especially where a 'climate-proofed' approach is adopted, mainstreaming adaptation into development aid could undo progress made against the principles of country ownership and public participation (Klein, 2008). This is because in targeting mainstreaming into development cooperation portfolios, rather than developing-country processes, responsibility rests with donor agencies rather than with the domestic institutions.

Instead, this paper proposes that focusing on developing country institutions and processes is likely to encourage a more holistic and integrated approach to adaptation mainstreaming, because by definition, the enabling environment of development investments is also taken into consideration. As such, this paper proposes a definition of climate change adaptation mainstreaming that is based on that for environmental mainstreaming put forward by Dalal-Clayton and Bass (2009), and in line with the definition proposed by Gupta and Van Der Grijp (2010):

> Mainstreaming should result in the informed inclusion of relevant climate vulnerability concerns into the decisions and institutions that drive national, sectoral, and local development policy, rules, plans, investment and action. This can be achieved in part through development cooperation – and mainstreaming adaptation into donor portfolios would be part of the alignment process – but the target of mainstreaming is national and sub-national level processes, and the key agents of mainstreaming are national and sub-national government and non-government stakeholders.

3.3. *A framework for mainstreaming*

The need for developing countries to mainstream adaptation into development planning is reflected in various avenues under the UNFCCC. Article 4.1 of the UNFCCC calls for Parties to take climate change adaptation into account in development planning. Guidance for the development of National Adaptation Programmes of Action (NAPAs) under the UNFCCC states that NAPAs should be 'mainstreamed' into national development planning processes (LEG, 2002). Various guidance exists on 'how to mainstream' adaptation into development, but these are generally targeted 'how-to' guides aimed at development professionals (OECD, 2009; UNDP-UNEP, 2011). Some early guidance was developed for mainstreaming NAPAs into development planning (LEG, 2002), but this was annexed in the overall NAPA development guidelines, and given limited funds for NAPA preparation, many countries did not have the resources or incentives to ensure an integrated approach to NAPA development (Burton & Van Aalst, 2004).

Huq and Ayers (2008) propose a framework for mainstreaming at the national level (see Figure 1). As with

Four Steps to National Capacity Building on Climate Change

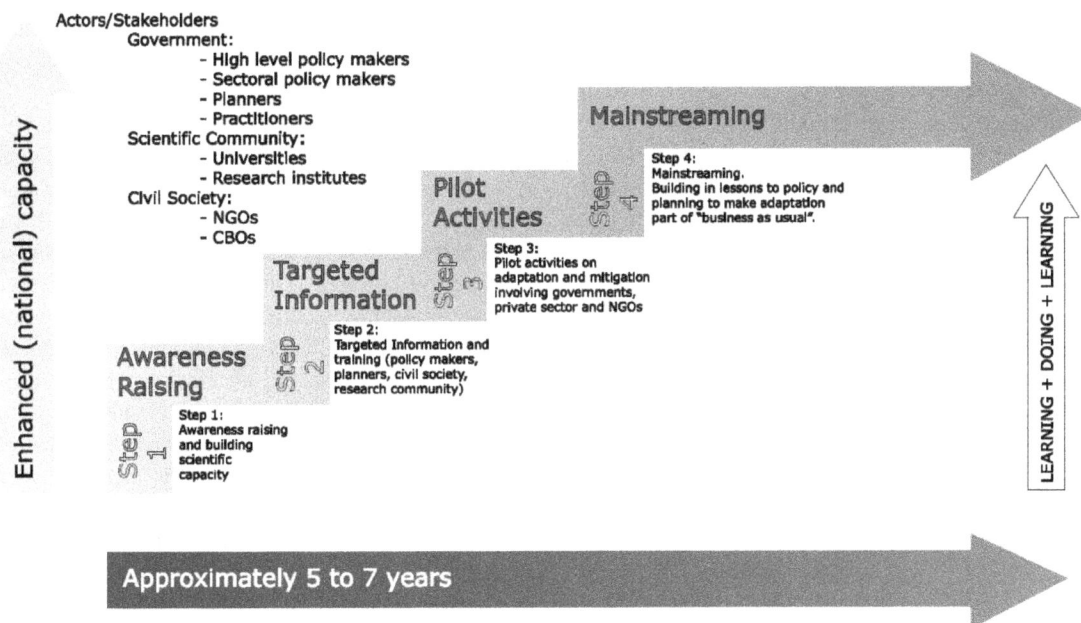

Figure 1. Four steps to building national capacity on climate change adaptation for mainstreaming.
Source: Huq and Ayers (2008).

other guidance developed at this time, the context of the framework assumed the drive for mainstreaming would come largely through international cooperation, given that incentives for climate change adaptation planning at the time were generally externally driven (Ayers & Huq, 2009a). This paper revisits this framework in the context of Bangladesh in the light of experiences around mainstreaming since 2008.

The framework proposes a linear sequence of awareness and scientific capacity-building, targeted information and training of key stakeholders, which is followed up with pilot studies to inform policy-makers and generate incentives to incorporate lessons learnt into policy and planning (Lebel et al., 2012) (see Figure 1).

Step one describes awareness-raising on the relevance of climate change adaptation for development. This is the first step in any group of decision-makers adopting adaptation as a priority issue. If adaptation is to be integrated into planning in a sustainable way, demand needs to come from policy planners and implementers themselves, requiring decision-makers to recognize adaptation as not only applicable, but, in some cases, urgent.

The authors argue that critical to getting adaptation to be taken up by policy-makers is the generation of scientific evidence to support decision-making. Simply highlighting 'problems' is not useful for policy-making; evidence generated needs to demonstrate relevant, realistic, solutions (Anderson, Ayers, & Kaur, 2011). Information that is generated in-country is more likely to be relevant to the

decision-making contexts of country decision-makers (Huq & Ayers, 2008).

This is challenging where impacts of climate change are highly uncertain, with uncertainty in climate change projections, and also because complex interactions between climate change and the social-development context determine how impacts will manifest. Much information around climate change impacts exists externally, in the realm of international bodies such as the IPCC. Thus, a first step is to invest in national-level capacity to generate locally appropriate evidence that can speak to policy decision-making forums. Supporting this step requires harnessing national-level expertise around vulnerability as well as building capacity around climate science, to ensure that adaptation priorities are country-owned and nationally responsive.

Step two describes how this information is made available to decision-makers across sectors and scales. First, there needs to be enough interest from decision-makers to demand and be receptive to climate vulnerability information. Second, information needs to be presented in a useable form, and capacity needs to be built to enable its use. Civil society plays a key 'boundary organization' role in translating scientific information into usable policy advice.

Step three describes the initial types of climate change adaptation responses, which tend to be isolated pilots and projectized interventions, often undertaken by NGOs. There has been criticism of the ways projectized adaptation approaches fail to lead to long-term resilience-building (Boyd, Grist, Juhola, & Nelson, 2009; Dodman & Mitlin, 2011; Schipper, 2007). Schipper (2007) suggests that in a projectized approach, adaptation is automatically taken as an objective or outcome, rather than as a process. Adaptation as a 'process' involves building adaptive capacity by creating enabling conditions for adaptive activities to take place. Nevertheless, this step has proved important for countries to learn about what adaptation might 'look like', to inform mainstreaming and build capacity.

Step four involves full integration of climate change adaptation into policy and planning across different sectors and scales, requiring a shift from 'business as usual' to investment and planning that is not only climate-proof, but also explicitly seeks to build resilience amongst the climate-vulnerable poor. It is this stage where Government stakeholders become fully engaged in adaptation planning. Critically, this means not just environment agencies, but planning and finance ministries who can drive integration of climate change adaptation priorities into broader development priorities.

4. The need for mainstreaming adaptation into development in Bangladesh

Bangladesh is frequently cited as one of the most vulnerable countries to climate change (Huq, 2001; Huq & Ayers, 2008; Rahman & Alam, 2003; UNDP, 2007) both because its geography makes it physically exposed to climatic hazards and also because of the socio-economic factors that make people vulnerable. Social vulnerability often drives physical exposure, which in turn can exacerbate social vulnerabilities.

Located on the Bay of Bengal with a flat and low-lying topography, Bangladesh is exposed to major storm and cyclones as well as flooding. Most of Bangladesh is less than 10 m above sea level, with almost 10% of the country below 1 m. Between 1960 and 2002, Bangladesh experienced over 40 cyclones with up to half a million human casualties per event (Huq & Khan, 2006).

Further, Bangladesh is one of the largest deltas in the world, formed by a dense network of the tributaries of the Ganges, Brahmaputra and Meghna Rivers. Eighty per cent of land is floodplain, so the majority of Bangladesh (with the exception of the far west 'highlands') is prone to flooding at least part of the year (MOEF, 2005). Many of the projected impacts of climate change are expected to exacerbate these existing environmental hazards: increasing intensity of cyclones and extreme events; exacerbating flooding and salinity intrusion.

The development characteristics of Bangladesh make it particularly vulnerable and limit adaptive capacity. Bangladesh is defined as one of the 'Least Developed Countries', with a GDP per capita (PPP US$) of 1241; a life expectancy at birth of 67.5 years; and an adult literacy rate of 53.5%

(UNDP, 2009). Furthermore, Bangladesh is particularly vulnerable due to dependency on agriculture. Two-thirds of the population is engaged in farming (although more than three-quarters of export earnings come from the garment industry) (Huq & Ayers, 2008).

Everyone in Bangladesh is not equally vulnerable to climate change. Reid and Simms (2007) suggest that the urban poor are especially vulnerable, because of the fragility of infrastructure in slums, and lack of employment security. In rural areas, those with insecure land tenure, particularly the lower Adivasi castes, are particularly vulnerable. Inherent gender inequalities in various social, economic and political institutions make women particularly vulnerable. Land access is particularly problematic for women because it is often obtained on a limited usufruct basis through marriage, which can leave women landless on divorce, and denies them collateral (Reid & Simms, 2007).

The combination of physical and social vulnerability means that in Bangladesh, climate change adaptation and development must be tackled together. Managing physical climate hazards without also addressing factors related to underdevelopment means that people would remain vulnerable. Only addressing development without taking into account existing and anticipated climate hazards means that development interventions are likely to prove unsustainable and possibly maladaptive in the long term.

5. Progress towards mainstreaming in Bangladesh

Bangladesh has approached adaptation mainstreaming both from a climate change perspective, through development of climate change specific plans, programmes and institutions that address developmental aspects of vulnerability, and also from a development perspective, integrating climate risk into development programmes and policies to help build broader cross-sectoral resilience.

In terms of climate-specific planning, Bangladesh was one of the first countries to develop its NAPA, in 2005. The NAPA proposed 15 projects that would contribute towards meeting Bangladesh's 'urgent and immediate' adaptation needs (MOEF, 2005). To date, one NAPA project has gone forward for implementation with funding from the Least Developed Countries (LDCs) Fund: 'Coastal Community-Based Adaptation to Climate Change through Coastal Afforestation in Bangladesh' (MOEF, 2008). The NAPA was updated in 2009, presenting nine short-term projects and nine medium-term projects. Although the NAPA is generally well regarded, it has faced criticism for adopting a relatively 'stand-alone' approach to adaptation through targeted climate change adaptation projects. Further, the process was developed in response to international policy and financial incentives under the UNFCCC, rather than being a product of national political will (COWI/IIED, 2009).

The National Capacity Self-Assessment (NCSA) for implementing the provisions of multilateral agreements, including the UNFCCC, was launched in 2007, and capacity-building for climate change received high priority. The Capacity Development Action Plan of NCSA identified a package of 15 actions for climate change, including capacity-building of relevant ministries and agencies for adaptation and mitigation.

Following the NAPA, the Government of Bangladesh, with support from development partners including the UK DFID, instigated the Bangladesh Climate Change Strategy and Action Plan (BCCSAP). Updated in 2009, the BCCSAP is now the main national planning document for climate change action in Bangladesh. The BCCSAP is widely regarded as having built on progress made under the NAPA, taking forward the research and recommendations into a more integrated and strategic planning framework. The BCCSAP is a 'pro-poor' climate change management strategy which prioritizes adaptation and disaster risk reduction, and also addresses low carbon development, mitigation, technology transfer and mobilization of international finance. The BCCSAP (MOEF, 2009) has six pillars:

(i) Food security, social protection and health,
(ii) Comprehensive disaster management,
(iii) Infrastructure,
(iv) Research and knowledge management,
(v) Mitigation and low carbon development,
(vi) Capacity-building and institutional strengthening.

There are 44 programmes under the BCCSAP. A 2.5 million USD Technical Assistance programme is being implemented by ADB to support BCCSAP implementation, including capacity-building of the Ministry of Environment and Forests (MOEFs) as well as other ministries involved in implementation.

There are two main trust funds to support implementation of the BCCSAP. One is funded by the Government of Bangladesh – the Bangladesh Climate Change Trust Fund (BCCTF), at a size of 100 million USD. More than 100 projects have been approved under the BCCTF (Pervin, 2013). The second is funded by several donors, the Bangladesh Climate Change Resilience Fund (BCCRF), with over 170 million USD to date. This dual approach is a resolution resulting from tensions over fund management control between the Government of Bangladesh and international agencies concerning fiduciary risk (Hedger, 2011). Projects submitted to either fund must conform to the needs and priorities identified in the BCCSAP.

Figure 2 presents the institutional arrangements supporting climate change in Bangladesh. The 2010 Climate Change Trust Fund Act provides guidance on how BCCTF funds can be disbursed and the supporting national institutional arrangements. The Climate Change Act established:

- A Technical Committee, chaired by the Secretary of MOEF with multi-stakeholder membership including from civil society. The Technical Committee reviews and evaluates project proposals to the National Trust Fund. There are subcommittees with key experts related to each pillar of the BCCSAP.
- The Trust board, which has the ultimate decision on applications to the BCCTF. Membership comprises 10 ministries and 17 members. The Technical Committee makes recommendations to the Trust Board, which often then requests further information before making a decision.

The MOEF is the focal ministry providing coordination and technical leadership on climate change, having led development of both the NAPA and BCCSAP. MOEF is considering the creation of a Department of Climate Change. There are Climate Change Cells within each ministry to mainstream climate change across all sectors. A Climate Change Unit has been established under MOEF to coordinate the various Climate Change Cells and build capacity across ministries.

An All-Party Parliamentary Group (APG) on Climate Change and Environment was established in 2009. It is a cluster of 121 MPs and is the largest APG representing all major Parties. Non-government institutions also play a key role in both climate risk management planning and implementation. For example, the working groups responsible for preparing the BCCSAP and NAPA had membership from and in some cases were led by national NGOs. National NGOs will also play the role of implementing entities under the BCCRF and BCCTF.

Climate change adaptation is also being integrated into general development planning. Vision 2021 and the National Perspective Plan set the development targets for Bangladesh up to 2021. Vision 2021 lays down a development scenario where citizens will have a higher standard of living, with better education, improved social justice, a more equitable socio-economic environment; and sustainability of development will be ensured through better protection from climate change and natural disasters. Implementation of Vision 2021 will be done through two medium-term development plans, the first (the sixth five-year plan) spanning 2011–2015. All three documents – Vision 21, the National Perspective Plan and the 6th 5-Year Plan – have chapters on climate change. The National Planning Commission is integrating climate change into the Annual Development Programme, which involves mainstreaming climate change into 28 projects in four sectors: agriculture, transport, rural development and water (IIED, 2011). The Planning Commission is currently reviewing

Conceptual Framework on Climate Change Related Policy and Institutions in Bangladesh

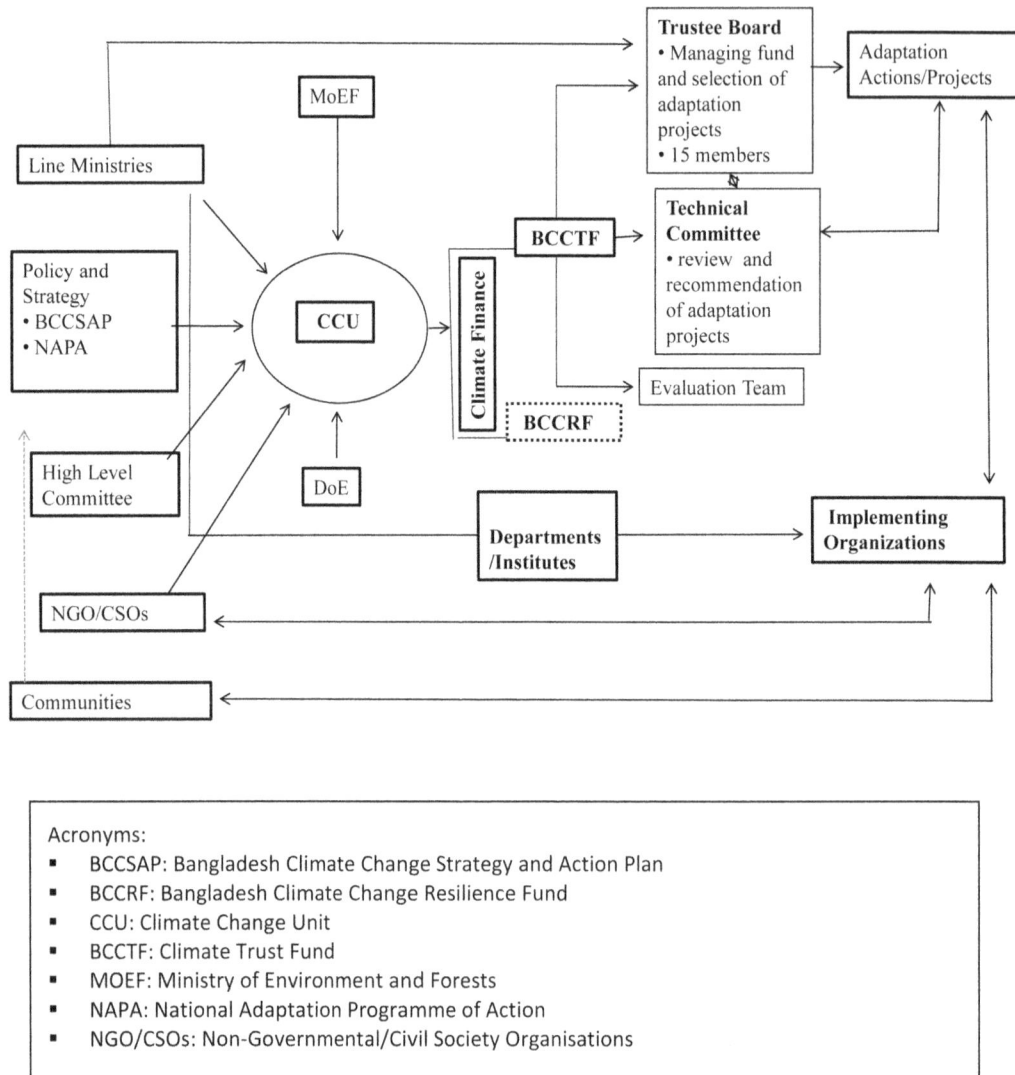

Figure 2.   Conceptual framework on climate change-related policy and institutions in Bangladesh.
Source: Huq and Rabbani (2011).

all existing policies from ministries to assess whether they conflict with climate change issues, and will then advocate changes if conflicts are identified (Pervin, 2013).

Climate change has also been integrated across relevant sectors in Bangladesh. In agriculture, climate risks are highlighted in agricultural planning documents including the National Agricultural Policy (2010). Bangladesh also leads the way on agricultural research programmes related to drought and saline-tolerant rice varieties, seen as key adaptation options (Agrawala et al., 2003; IIED, 2011). Recommendations from the World Bank on the impacts of climate change have been incorporated into coastal zone management programmes, in the preparation of disaster preparedness plans and in the 25-year water sector plan. Climate change is recognized by the National Water Management Plan (2001) as one of the factors determining future water

management. Many of the Plan priorities are synergistic with adaptation, such as the recommendation for early warning and flood-proofing systems.

6.   Lessons from Bangladesh: progress against the four steps

How do the experiences above relate to the framework for mainstreaming? What can we learn about the 'four steps' proposed?

First, the experience of Bangladesh demonstrates significant progress against each of the four steps. Under step one of using scientific capacity and knowledge generation, Bangladesh has built up a significant body of national-level expertise around climate change and adaptation options. Bangladesh has a relatively long history of

10

engagement in climate change studies and adaptation interventions compared to other LDCs (Ayers & Huq, 2009b). As a result, there are a large number of organizations and agencies with knowledge, tools and capacity to assess climate-related impacts. In addition, Bangladesh has also long been the 'face of climate vulnerability' to the international community. This, coupled with its long history of engagement in international climate change fora, has made Bangladesh the focus of many international studies on climate change impacts (ibid). Therefore climate data and expertise available for and in Bangladesh are considerable and growing. For example, research being conducted in Bangladesh around drought-, flood-, and saline-tolerant rice varieties is considered cutting-edge globally (IIED, 2011).

In relation to step 2, targeted information-sharing, training and capacity-building have been central to almost all of the climate change planning activities and investments in Bangladesh. For example, the NAPA involved several cross-country workshops at sub-national level to raise awareness on climate impacts. The ADB is working with the Government of Bangladesh as part of the Pilot Programme for Climate Resilience (PPCR) on a significant capacity-building and knowledge management programme. This supports generation, dissemination and application of information and knowledge products as the means to influence policies and address the impacts of climate change. It intends to result in improved knowledge management systems and institutions, and enhanced capacities of government institutions to make climate-resilient decisions (climateinvestmentfunds.org, 2011). Bangladesh is also home to the International Centre for Climate Change and Development (ICCCAD), which conducts training for government and non-government stakeholders on climate and development issues. In 2011, ICCCAD conducted a course for international Government stakeholders, including several from Bangladesh, on climate change mainstreaming (see centers.iub.edu.bd/icccad).

There has also been significant experience around pilot and projectized interventions in Bangladesh in line with step 3. The Bangladesh NAPA process identified 18 projects for short- and medium-term investment (see above). The BCCRF and BCCTF have now approved projects towards implementation of the BCCSAP. Bangladesh also has a significant number of community-based adaptation (CBA) projects largely implemented by NGOs. Learning from these projects for informing wider scale adaptation planning has been promoted through engagement in the International CBA Conferences which have taken place annually (Bangladesh hosts biannually) and which increasingly attract government stakeholders (IIED, 2013).

These projectized approaches to adaptation have faced some criticism for not sufficiently engaging with longer term policy and institutional frameworks that would enable sustainable mainstreaming of climate change

adaptation at higher levels (Ayers, 2011; Dodman & Mitlin, 2011). For example, a review undertaken by COWI/IIED of the Bangladesh NAPA in 2009 suggested that there were inadequate mechanisms for comprehensive multistakeholder participation; and a capacity deficit to manage adaptation projects and investments (COWI/IIED, 2009). This echoes more general criticisms of project-based approaches to adaptation. For example, Schipper (2007) suggests that in taking a projectized approach to adaptation, adaptation is automatically taken as an objective or outcome, rather than as a process. This contradicts a vulnerability-based perspective on adaptation, which involves a process of building adaptive capacity by creating the enabling conditions for adaptation to take place. Indeed, the notion of meeting 'urgent and immediate' needs reveals that adaptation is something that can be done in the short term, and not part of a longer term planning process. As noted by Schipper, from a vulnerability perspective,

> Adaptation to climate change is not as simple as designing projects, drawing up a list of possible adaptation measures and implementing these. It requires a solid development process that will ensure that the factors that create vulnerability are addressed. (Schipper, 2007, p. 6)

However, Bangladesh has done much to move beyond an isolated approach to adaptation project planning towards step 4 in the framework. For example, the wider knowledge generation and capacity-building benefits of undertaking the NAPA process contributed to both the political will and capacity underpinning the development of the nationally driven BCCSAP climate change funding streams (COWI/IIED, 2009). The CBA conference in 2013, hosted in Dhaka, focused on 'mainstreaming CBA', outlining the need to integrate projectized approaches into existing planning systems (IIED, 2013), while the 2011 CBA Conference, also in Dhaka, focused on 'Upscaling CBA'.

Finally, the BCCSAP is widely regarded as a comprehensive and integrated example of adaptation planning. The Plan itself has elements of 'climate-proofing', but also explicitly recognizes the need for a more integrated, development-first approach to adaptation planning. For example, the plan does not only look at the impacts of climate change on agriculture, but also the role of agriculture and food security in building longer term adaptive capacity (MOEF, 2009). This is exemplified by the prioritization of pillars of social protection and health. The plan intends to cut across all sectors, with 44 programmes so far identified within six thematic areas, including emphasis on strengthening human resources and institutional capacity (MOEF, 2009). The fact the plan was nationally driven and is being funded in part by national funding demonstrates the political will and ownership of the Government to managing climate risks in an integrated way.

Further, the integration of climate change adaptation in national and sectoral development strategies demonstrates commitment to ensuring that development is both 'climate-resilient' and also builds climate resilience. Climate resilience is being integrated into 'business as usual' planning systems.

7. Moving beyond the four steps

Despite this progress, feedback from those engaged in climate change planning in Bangladesh revealed many challenges in implementing these four steps. Challenges listed by interviewees included inadequate coordination mechanisms among various ministries and line agencies, limited coordination capacity of the MOEF and other implementing agencies, losses of institutional memory in relevant agencies and 'brain drain' of trained officials, leading to delays in knowledge generation and maintenance. Projects often emerged in an ad hoc manner, and programme cycles were affected by interruptions in the flow of climate funds, delays in procurement and disbursement, unavailability of qualified staff, and the overburdening of the limited number of technical staff that did exist. Finally, full integration of climate change into national development planning is entirely dependent on strong political commitment. In Bangladesh, interviewees suggested that a turbulent

political system with frequent changes of the ruling party manifesto or development agenda has led to erratic and unpredictable progress in mainstreaming adaptation.

These challenges suggest the need to move beyond this four-step model of mainstreaming climate change. Firstly, experiences suggest that the process of mainstreaming is not linear, with each step building on the last. For example, while undertaking adaptation projects did result in generation of knowledge and capacity that could be built under the BCCSAP, projects continue to be implemented alongside more integrated approaches and do hold value in their own right. Furthermore, the line between 'projects' and 'mainstreamed plans' is not distinct, as projects themselves can be mainstreamed into existing planning processes. The Community Climate Change Programme (CCCP) under the BCCRF supports CBA projects as outlined in the BCCSAP, implemented by NGOs (CCCP, 2013).Thus, while Bangladesh reflects all four proposed 'steps' towards mainstreaming adaptation, it also shows that, in practice, the pathway to mainstreaming is not linear. It is made up of a patchwork of processes, stakeholders and approaches that converge or coexist.

Second, while information or evidence is often perceived as a prerequisite for decision-making, experience in Bangladesh demonstrates that a lot of decision-making around climate change adaptation takes place in the face of uncertainty. A study by IIED on climate change decision-making in Bangladesh showed that where

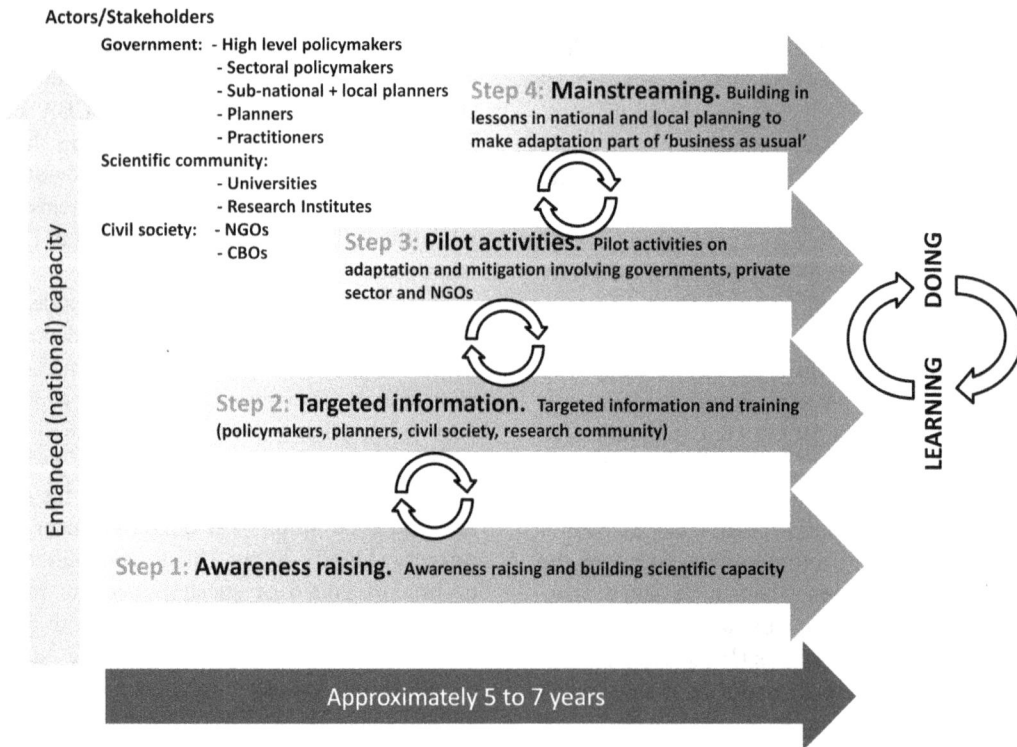

Figure 3. Revised framework for building national capacity on climate change adaptation for mainstreaming.

evidence on climate change impacts was lacking, alternative sources of information can be used stemming from 'non-scientific' arenas, such as community-based knowledge about climate trends and adaptive responses. Further, most decision-making is not neutrally 'evidence-based', but based on a complex set of political drivers that influence not only the way information is used, but also how that information is generated in the first place (Anderson et al., 2011; IIED, 2011).

Third, Bangladesh is showing that information around climate change can be generated from a diverse set of stakeholders, not necessarily only 'climate experts'. Extensive consultations were involved in development of the BCCSAP with a variety of stakeholder groups, especially development planners and cross-sectoral experts. Further, the BCCSAP is taken as a 'living document' – it can be updated in response to new information and emerging priorities.

Furthermore, while the four-step framework emphasizes national-level institutions, it could perhaps be enhanced by elaborating in the final step how adaptation is taken up by planners and decision-makers at sub-national and local levels to achieve effective national mainstreaming of adaptation (see Figure 3). In order to effectively build the resilience of the climate-vulnerable poor at local levels, there is further need for capacity-building of local-level institutions (Christensen et al., 2012).

8. Conclusions

Given the links between adaptation and development, it makes sense to address the two in an integrated way, through 'mainstreaming'. But mainstreaming has many different interpretations – it can mean integrating adaption into development planning, or development into adaptation planning. It can mean managing climate risks in to development, or finding more transformational ways of dealing with vulnerability to both climate and other risks. There are also a number of options for operationalizing the concept, depending on how mainstreaming is interpreted, and also what the target of planning is – whether we think of 'development' as international development investments, or national development planning strategies and budgets. This paper has suggested that for mainstreaming to be sustainable, the object of mainstreaming should be national and sub-national institutions and processes. It has also shown that there is no single 'best' approach to doing mainstreaming – mainstreaming emerges as a patchwork of climate-proofing and more integrated strategies that all contribute to building climate resilience in interconnected ways.

This paper has also shown that the four-step framework for mainstreaming is useful in presenting some of the prerequisites for enabling mainstreaming to take place in principle. However, in practice, we have shown that

mainstreaming is not linear (see revised Figure 3). In Bangladesh, mainstreaming has emerged in a number of different guises, all involving a blend of information, capacity building, resource-mobilization and governance changes, underpinned by political will. Different sets of stakeholders have been engaged, from both Government and non-government sectors, and within these groups, some have moved faster than others. This is perhaps no great surprise, and indeed other scholars have pointed to the role that different policy stakeholders can play in driving differential policy change processes (see Huitema & Meijerink, 2010).

In the light of the experience of Bangladesh, we therefore suggest that while the four-step framework could be used alongside the various other 'how to' guides in understanding some of the activities that mainstreaming entails, we recommend further research into the conditions that give rise to effective mainstreaming in different contexts. For example, while projectized approaches undertaken during the NAPA and other CBA activities in Bangladesh were indeed early stage and perhaps not sustainable in their own right, they led to the building of experience, expertise and, critically, agency that enabled more integrated planning through the BCCSAP. Such experiences built formal and informal networks of subnational, national and international stakeholders across different agencies that later came together to promote integrated national planning. Further work on capturing the matrix of activity in different countries that builds the will, momentum, expertise and networks to achieve integrated planning is critical for informing future sustainable mainstreaming. Once identified, work to strengthen such preconditions is critical, such as fostering national expertise through training, and nurturing networks through formalized deliberative platforms and conferences. Linear models and 'how-to' frameworks can only be useful as a way of guiding, or perhaps reviewing, what is essentially a messy organic process of integrated policy-making. Instead, further efforts should focus on identifying and cultivating the roots of effective mainstreaming.

References

Adger, N., Huq, S., Brown, K., Conway, D., & Hulme, M. (2009). Adaptation to climate change in the developing world. In L. Schipper & I. Burton (Eds.), *The earthscan reader on adaptation to climate change* (pp. 295–312). London: Earthscan.

Adger, N., & Kelly, P.M. (1999). Social vulnerability to climate change and the architecture of entitlements. *Mitigation and Adaptation Strategies to Global Change, 4*, 253–266.

Agrawala, S., Ota, T., Uddin Ahmed, A., Smith, J., & van Aalst, M. (2003). *Development and climate change in Bangladesh: Focus on coastal flooding and the Sundarbans*. Paris: OECD.

Anderson, A. (2011). Climate change and poverty reduction. Climate and development knowledge network Policy Briefing, August 2011. Retrieved from http://cdkn.org/resource/climate-change-and-poverty-reduction/?loclang=en_gb

Anderson, S., Ayers, J., & Kaur, N. (2011). *Evidence paper 1 prepared for the South Asia Climate Resilience Alliance scoping phase.* Unpublished, IIED, London.

Ayers, J. (2011). Resolving the adaptation paradox: Exploring the potential for deliberative adaptation policy-making in Bangladesh. *Global Environmental Politics, 11*(1), 62–88.

Ayers, J., & Dodman, D. (2010). Climate change adaptation and development I: The state of the debate. *Progress in Development Studies, 10*(2), 161–168.

Ayers, J., & Huq, S. (2009a). Supporting adaptation to climate change: What role for ODA? *Development Policy Review 27*(6), 675–692.

Ayers, J., & Huq, S. (2009b). The value of linking mitigation and adaptation: A case study of Bangladesh. *Environmental Management, 43*(5), 753–764.

Ayers, J., Kaur, N., & Anderson, S. (2011). Negotiating climate resilience in Nepal. *IDS Bulletin, 42*(3), 70–79.

Barnett, J., & O'Neill, S. (2010). Maladaptation. *Global Environmental Change, 20*(2), 211–213.

Blaikie, P.M., Cannon, T., Davis, I., & Wisner, B. (1994). *At risk: Natural Hazards, people's vulnerability and disasters.* London: Routledge.

Booth, C., & Bennett, C. (2002). Gender mainstreaming in the European Union: Towards a new conception and practice of equal opportunities? *European Journal of Women's Studies, 9*(4), 430–46.

Boyd, E., Grist, N., Juhola, S., & Nelson, V. (2009). Exploring development futures in a changing climate: Frontiers for development policy and practice. *Development Policy Review, 27*(6), 659–674.

Burton, I. (2004). *Climate change and the adaptation deficit. Adaptation and Impacts Research Group Occasional Paper 1.* Quebec: Environment Canada.

Burton, I., Diringer, E., & Smith, J. (2006). *Adaptation to climate change: International policy options.* Arlington, VA: PEW Centre on Global Climate Change.

Burton, I., Huq, S., Lim, B., Pilifosova, O., & Schipper, E.L. (2002). From impacts assessment to adaptation priorities: The shaping of adaptation policy. *Climate Policy, 2*(2–3), 145–159.

Burton, I., & Van Aalst, M. (2004). *Look before you leap: A risk management approach for incorporating climate change adaptation in world bank operations.* Washington, DC: World Bank.

Cannon, T. (2000). Vulnerability analysis and disasters. In D.J. Parker (Ed.), *Floods* (pp. 43–55). London: Routledge.

CCCP. (2013). *Community climate change project website.* Retrieved from http://www.pksf-cccp-bd.org/

Christensen, K., Raihan, S., Ahsan, R., Uddin, A.M.N., Ahmed, C.S., & Wright, H. (2012). *Financing local adaptation: Ensuring access for the climate vulnerable in Bangladesh.* Dhaka: ActionAid Bangladesh, ARCAB, BCAS, and ICCCAD. Retrieved from http://www.actionaid.org/sites/files/actionaid/financing_local_adaptation.pdf

COWI/IIED. (2009). *Evaluation of the operation of the least developed countries fund for adaptation to climate change.* Copenhagen: Ministry of Foreign Affairs of Denmark.

Dalal-Clayton, B., & Bass, S. (2009). *A Guide to Environmental Mainstreaming (Rough First Draft).* IIED, UK. Retrieved from www.environmental-mainstreaming.org/

Dodman, D., Ayers, J., & Huq, S. (2009). Building resilience. In Worldwatch Institute, 2009 (Ed.), *State of the world 2009: Into a warming world* (pp. 151–168). Washington, DC: Worldwatch Institute.

Dodman, D., & Mitlin, D. (2011, February). Challenges to community-based adaptation. *Journal of International Development.* doi:10.1002/jid.1772

Ford, J. (2008). Emerging trends in climate change policy: The role of adaptation. *International Public Policy Review, 3*(2), 5–16.

Gupta, J. (2009). Climate change and development cooperation: Trends and questions. *Current Opinion in Environmental Sustainability, 1*(2), 207–213.

Gupta, J., & Van Der Grijp, N. (Eds.). (2010). *Mainstreaming climate change in development cooperation: Theory, practice and implications for the European Union* (pp. 303–341, 1st ed.). Cambridge: Cambridge University Press.

Hedger, M. (2011). *Climate finance in Bangladesh: Lessons or development cooperation and climate finance at the national level. EDC 2020 Policy Brief.* Retrieved from http://www.edc2020.eu/fileadmin/publications/EDC_2020_-_Policy_Brief_no_14_-_Climate_Finance_in_Bangladesh.pdf

Huitema, D., & Meijerink, S. (2010). Realizing water transitions the role of policy entrepreneurs in water policy change. *Ecology and Society, 15*(2), 26–36.

Huq, S. (2001). Climate change and Bangladesh. *Science, 294,* 1617–1617.

Huq, S., & Ayers, J. (2007). *Critical List: The 100 nations most vulnerable to climate change. Sustainable development opinion.* London: IIED.

Huq, S., & Ayers, J. (2008). Streamlining adaptation to climate change into development projects at the national and local level. In European Parliament (Ed.), *Financing climate change policies in developing countries* (pp. 52–68). Brussels: European Parliament.

Huq, S., & Khan, M. (2006). Equity in national adaptation programs of action. In N. Adger, J. Paavola, S. Huq, & M.J. Mace (Ed.), *Fairness in adaptation to climate change* (pp. 131–153). Cambridge: MIT Press.

Huq, S., & Rabbani, G. (2011). Climate change and Bangladesh: Policy and institutional development to reduce vulnerability. *Journal of Bangladesh Studies, 13*(1), 1–10.

Huq, S., Reid, R., Konate, M., Rahman, A., Sokona, Y., & Crick, F. (2004). Mainstreaming adaptation to climate change in least developed countries (LDCs). *Climate Policy, 4*(1), 25–43.

IIED. (2011). *Policy discourse analysis report, Bangladesh. Unpublished scoping study to inform the design of the South Asia Alliance on Climate Resilience (SACRA).* London: IIED.

IIED. (2013). *Community based adaptation: Mainstreaming CBA into national and local planning.* IIED and BCAS. Retrieved from http://pubs.iied.org/pdfs/G03616.pdf?

IPCC. (2007). Summary for policymakers. In M.L. Parry, O.F. Canziani, J.P. Palutikof, P.J. van der Linden, & C.E. Hanson (Eds.), *Climate change 2007: Impacts, adaptation and vulnerability. Contribution of working group II to the fourth assessment report of the intergovernmental panel on climate change* (pp. 7–22). Cambridge: Cambridge University Press.

Kelly, P.M., & Adger, W.N. (2009). Theory and practice in assessing vulnerability to climate change and facilitating adaptation. In L. Schipper & I. Burton, (Eds.), *The earthscan reader on adaptation to climate change* (pp. 161–186). London: Earthscan.

Klein, R.J.T. (2008). Mainstreaming climate adaptation into development policies and programmes: A European perspective. In European Parliament (Ed.), *Financing climate change policies in developing countries* (pp. 38–50). Brussels: European Parliament.

Klein, R.J.T. (2010). Mainstreaming climate adaptation into development: A policy dilemma. In A. Ansohn & B. Pleskovic

(Ed.), *Climate governance and development* (pp. 35–33). Washington, DC: World Bank.

Klein, R.J.T., Eriksen, S.E.H., Naess, L.O., Hammill, A., Tanner, T.M., Robledo, C., & O'Brien, K.L. (2007). Portfolio screening to support the mainstreaming of adaptation to climate change into development assistance. *Climatic Change, 84* (1), 23–44. doi: 10.1007/s10584-007-9268-x

Klein, R.J.T., Schipper, L., & Dessai, S. (2003). *Integrating mitigation and adaptation into climate and development policy: Three research questions.* Tyndall Centre Working Paper 405. Tyndall Centre.

Lebel, L., Li, L., & Krittasudthacheewa, C. (2012). *Mainstreaming climate change adaptation into development planning.* Bangkok: Adaptation Knowledge Platform and Stockholm Environment Institute, pp 32.

LEG. (2002). *Annotated guidelines for the preparation of national adaptation programmes of action.* Least Developed Countries Expert Group, Bonn: UNFCCC.

Levina, E. (2007). *Adaptation to climate change: International agreements for local needs.* Document prepared by the OECD and IEA for the Annex I Expert Group on the UNFCCC. Paris: OECD/IEA.

McGray, H., Hammill, A., & Bradley, R. (2007). *Weathering the storm: Options for framing adaptation and development.* Washington, DC: World Resources Institute.

Ministry of Environment and Forest Government of the People's Republic of Bangladesh (MOEF). (2008). *NAPA project document: Community-based adaptation to .Climate Change through coastal afforestation in Bangladesh.* Project ID: PIMS 3873. Dhaka: MOEF/UNDP.

MOEF. (2005). *Bangladesh national adaptation programme of action (NAPA).* Bonn: UNFCCC, Ministry of Environment and Forest Government of Bangladesh.

MOEF. (2009). *Bangladesh climate change strategy and action plan.* Ministry of Environment and Forests, Government of Bangladesh. Retrieved from www.moef.gov.bd/climate_change_strategy2009.pdf

OECD. (2005). *The Paris declaration on aid effectiveness and the Accra agenda for action.* Paris: OECD.

OECD. (2009). *Integrating climate change adaptation into development co-operation: Policy guidance.* OECD Publishing. doi:10.1787/9789264054950-en

Olhoff, A., & Schaer, C. (2010). *Screening tools and guidelines to support the mainstreaming of climate change adaptation into development assistance – a stocktaking report.* New York, NY: UNDP.

Pervin, M. (2013). *Mainstreaming climate change resilience into development planning in Bangladesh.* Country report. London: IIED.

Rahman, A., & Alam, M. (2003). *Mainstreaming adaptation to climate change in least developed countries (LDCs).* IIED Working Paper2. London: IIED.

Reid, H., & Simms, A. (2007). *Up in smoke? Asia and the Pacific.* Up in Smoke Working Group on Climate Change and Development. London: New Economics Foundation.

Ross, A., & Dovers, S. (2008). Making the harder yards: Environmental policy integration in Australia. *Australian Journal of Public Administration, 67*(3), 245–260.

Schipper, L. (2006). Conceptual history of adaptation in the UNFCCC process. *RECIEL, 15*(1), 82–92.

Schipper, L. (2007). *Climate change adaptation and development: Exploring the linkages.* Tyndall Centre Working Paper Series 107. Tyndall Centre for Climate Change Research.

Seballos, F., & Kreft, S. (2011). Towards an understanding of the political economy of the PPCR. *IDS Bulletin, 42*(3), 33–41.

Sen, A.K. (1999). *Development as freedom.* Oxford: Oxford University Press.

Smit, B., & Wandel, J. (2006). Adaptation, adaptive capacity and vulnerability. *Global Environmental Change, 16*, 282–292.

Sperling, F. (2003). *Poverty and climate change: Reducing the vulnerability of the poor through adaptation.* Washington, DC: World Bank.

Tanner, T. (2008). *Climate risk management: Adapting development cooperation to climate change.* IDS Working Paper Wp2008–20.

UNDP. (2007). *Country-in-focus: Bangladesh.* UNDP-RCC web bulletin, 2.

UNDP. (2009). International human development indicators: Bangladesh. Retrieved April 2, 2013, from http://hdrstats.undp.org/en/countries/profiles/BGD.html

UNDP-UNEP. (2011). *Mainstreaming adaptation to climate change in development planning: A guidance for practitioners.* Nairobi, Kenya: UNDP/UNEP Poverty Environment Initiative.

Identifying operational mechanisms for mainstreaming community-based adaptation in Nepal

Bimal Raj Regmi and Cassandra Star

Discipline of Politics and Public Policy, School of Social and Policy Studies, Faculty of Social and Behavioural Sciences, Flinders University, Australia

Mainstreaming is a feasible and viable option for scaling up initiatives on community-based adaptation (CBA) to climate change. However, there is little evidence on how to get CBA mainstreaming feasible and to work effectively. This paper aims to investigate two major questions: (1) what kind of policies favour mainstreaming CBA; and (2) what kinds of approaches or practices are required to operationalize CBA mainstreaming in the case of Nepal? The field research for this paper was conducted in the Dhading, Nawalparasi and Pyuthan districts of Nepal. The research used a mix of approaches and methods for data generation and analysis. The findings reveal that policies to operationalize CBA mainstreaming should build on past policy successes and include community-centric provisions that empower local institutions and encourage them to practice inclusive decision-making and benefit-sharing mechanisms. One lesson from this analysis of the practices of mainstreaming in Nepal is that an integrated co-management approach to mainstreaming is necessary to overcome the barriers related to knowledge, finance and technology. It is concluded that the operational mechanisms of mainstreaming CBA in development should have an inclusive local structure and be responsive to national policies and governance arrangements.

1. Introduction and background

Community-based adaptation (CBA) to climate change, which has developed considerable currency with civil society organizations, is designed to help the poorest and most vulnerable adapt to climate change (Huq & Reid, 2007). It has often been referred to as a bottom-up adaptation approach, recognizing that the poorest and most vulnerable people should have direct access to the majority of finance for climate change adaptation, including technological support. CBA aims to build the resilience of communities by enhancing their capacity to cope and better adapt to both climate variability and changes (Ayers, Alam, & Huq, 2010). Despite rapid progress in the development and sharing of knowledge about CBA, challenges remain in expanding small, localized project responses to reach the wider communities. Reid, Huq, and Murray (2010) argue that whilst the number of CBA case studies is proliferating, it will be important to find practical ways to scale up the experiences and lessons to wider communities.

The literature has outlined that CBA is more likely to be effective if it is mainstreamed in national and local development (Ayers et al., 2010; Ayers & Huq, 2013). There is no universally accepted definition of mainstreaming; it has

been variously described. Agrawala (2005, p. 15) defines mainstreaming as 'the integration of climate change vulnerabilities or adaptation into some aspect of related government policy such as water management, disaster preparedness and emergency planning or land-use planning'. Similarly, Huq et al. (2004, pp. 35–36) define it as the integration of information, policies and measures to address climate change in ongoing developing planning and decision-making. Based on these definitions, mainstreaming in this paper is regarded to entail both policy responses as well as the practice of integrating climate change issues into regular development policy and planning, at all levels.

Analyses suggest that incorporating adaptation into mainstream development is a 'win-win' approach and that capitalizing on synergies in this way will lead to more efficient resource mobilization (Ayers & Huq, 2013; Huq, Reid, & Murray, 2006) and more sustainable, effective and efficient use of resources (Ayers & Huq, 2009; Huq & Ayers, 2008a; Huq et al., 2004, pp. 35–36). Similarly, other literature has outlined additional potential contributions of mainstreaming climate change adaptation, which may include avoidance of policy conflicts, reduction of risks and vulnerability, and promotion of individual/

organizational efficiency (Agrawala, 2004; Lebel et al., 2012; Persson, 2008; Srinivasan & Uchida, 2008). Specifically with regard to mainstreaming CBA, Ayers, Huq, Faisal, and Hussain (2013) argue that mainstreaming climate change adaptation will also contribute to the spirit of advancement in community-based initiatives as well as provide opportunities for effectively channelling adaptation financing and implementation. CBA is seen as an important strategy because it also helps to mainstream local-level adaptation innovations into development policy and practice (Reid & Schipper, 2014).

However, there is little evidence on how to ensure CBA mainstreaming works effectively. Debates about mainstreaming are dominated by issues related to policy and planning agendas, whilst in many countries the main challenge is to translate policy into action. This paper explores both these policy and implementation perspectives in relation to mainstreaming in Nepal. Such analysis is significant for Nepal and indeed other least developed countries (LDCs) as it will inform decision-making and planning regarding the best approaches for identifying operational mechanisms for mainstreaming CBA in development, and its wider scaling.

Nepal was selected as the case study in this paper because it is one of the most vulnerable LDCs due to its fragile landscape and poor socio-economic status (Ministry of Environment, 2011a). As with other LDCs, mainstreaming CBA to climate change is a priority for Nepal. The government of Nepal is making progress in devising policy to address climate change adaptation, mostly linking local, CBA practices with national level policies and plans. The experiences of Nepal are valuable for other LDCs seeking to mainstream CBA in national policies and programmes.

This study draws on an analysis of national level policy and empirical findings from the field. The policy analysis involves a review of major development and climate change policies of Nepal relevant to CBA mainstreaming. The empirical community research includes a field study focusing on the Dhading, Nawalparasi and Pyuthan districts of Nepal, selected because they were the pioneer districts involved in piloting the mainstreaming of climate change adaptation in Nepal. The experiences of these three districts are relevant to inform policy and practice on future up scaling of CBA practices in Nepal and other LDCs.

This paper argues that mainstreaming is a feasible and viable option for scaling up CBA initiatives so that they constitute more than just the sum of small, localized activities. However, mainstreaming CBA must be supported with integrated policies and efficient governance mechanisms that are accountable and responsive to vulnerable households and communities. Reporting on a research study, this paper investigates two major questions: (1) what kind of policies favour mainstreaming CBA; and (2) what

kinds of approaches or practices are required to operationalize CBA mainstreaming in the case of Nepal?

The following two sections outline the research methodology and discuss the findings in relation to these major research questions, the latter providing (a) information and evidence to justify the type of policy process and content required to mainstream CBA in development planning processes in Nepal; and (b) an analysis of the effectiveness of translating policies into practice, to examine how climate change adaptation is mainstreamed in this country.

2. Research methodology

This study involved primary research and was qualitative and exploratory in nature. Different methodological approaches were used to analyse policies and progress in their implementation. The methods for analysing policy built on existing literature relating to public policy analysis, namely the approaches proposed by Lasco et al. (2009) and Walt and Gilson (1994). Both approaches facilitate a systematic analysis of the many factors (content, process, context and actors) affecting climate change and development policy.

The advantage of using the model proposed by Walt and Gilson (1994) is that it analyses climate change and development policy from an actor-oriented perspective. The approach proposed by Lasco et al. (2009) used a mix of analyses in assessing how far climate change has been integrated in major development plans and programmes of the government: looking at policy and programme documents and interviewing people to map their perceptions. The advantage of both models is that they emphasize the views of those impacted by the policies.

The framework for mainstreaming adaptation into development planning proposed by Huq and Ayers (2008b) describes a linear sequence of building awareness and scientific capacity, targeting information, training key stakeholders, undertaking pilot studies to inform policymakers, and integrating 'learning' into 'doing' (Lebel et al., 2012). This framework is relevant for examining mainstreaming initiatives at the national level but it cannot be applied to the local context. For the current study a slight modification was made to look at how mainstreaming has contributed to: enhancing collaboration amongst agencies; building stakeholder capacity; strengthening the availability of information and knowledge; and institutionalizing/scaling up CBA.

There are prerequisites for successful CBA mainstreaming. Involving a diversity of actors and promoting a multi-stakeholder approach can help to address the complexities and uncertainties associated with climate change. Agrawal (2008, pp. 3–5) argues that the multifaceted nature of climate change demands institutional innovation and learning to forge partnerships and collaboration amongst diverse actors and agencies and to leverage the huge requirement for financial and technological resources. Similarly, Ridder and Team (2006) and Lobo (2011) argue that due to the complex nature of and uncertainty relating to climate change, stakeholders should learn together and forge strong alliances for knowledge and resource sharing.

This study examines progress in mainstreaming by looking at two specific case studies in Nepal. A case study approach was favoured because it supports in-depth analysis (Sjoberg, Williams, Vaughan, & Sjoberg, 1991). The two cases are the Poverty Environment Initiative (PEI) implemented by the Ministry of Local Development in Nepal with support from the United Nations Development Programme (UNDP); and the Local Adaptation Plan of Action (LAPA), which is seen as a means of integrating adaptation options into development policy and planning processes in Nepal and supporting local CBA practices. Local and CBA planning under the LAPA has been conducted in selected areas by the Livelihoods and Forestry Programme and national non-government organizations (NGOs). Whilst both case studies specifically tackle the mainstreaming of CBA, they differ notably in the fact that each is led by different agencies (government, donors or NGOs). Fieldwork for this research was conducted in the Jogimara, Sukrauli, Dhungegadi and Bangesal Village Development Committees (VDCs) in Nepal.

Jogimara VDC in Dhading district was selected as a case study because it is one of the first VDCs where a PEI has implemented mainstreaming of climate change in development planning. The second case study involved Sukrauli VDC of Nawalparasi and the Bangesaal and Dhungegadi VDCs of Pyuthan district were selected because they were the first three VDCs to carry out piloting of CBA mainstreaming through implementation of LAPA initiatives (see Figure 1 for the location of the study areas).

The data reported on in this paper were derived from four sources:

(1) secondary sources (government policies, plans and project documentation);
(2) opinions/perceptions of selected policy-makers and practitioners;
(3) opinions/perceptions of locals (local communities and community groups, for example, community forestry user group/farmers group); and
(4) field observations.

Between December 2011 and March 2012, 17 policy-makers were interviewed and one focus group discussion (FGD) was held. Eight FGDs were carried out involving more than 150 households and 7 community groups. Two FGDs were carried out in Jogimara, Sukrauli, Dhungegadi and Bangesaal VDCs. A total of 28 practitioners representing both government and non-government agencies were

Figure 1. Study districts.

interviewed using a semi-structured interview method. In addition, 3 FGDs were conducted with 7–10 practitioners in each.

For the semi-structured interviews, an open-ended questionnaire was used, focusing on aspects of: the policy-making process; the effectiveness of mainstreaming initiatives; and overall benefits of climate change interventions to the households and communities. FGDs involved a general discussion with households and communities in the study locations, and were used to map the perceptions of different categories of household and community members (based on ethnicity and gender) on the significance of mainstreaming CBA in local development planning.

Purposive sampling was used to select participants. This is a form of non-probability sampling in which decisions concerning the individuals to be included in the sample are taken by the researcher, based upon a variety of criteria (Oliver & Jupp, 2006). The criteria for participation used in this research included the possession of specialist knowledge and experience by policy-makers, practitioners and communities involved in climate change mainstreaming initiatives, and their willingness to participate in the research. The FGDs and interviews were focused on questions and issues relating to how climate change mainstreaming was supported in terms of policy and practice.

3. Major findings

3.1. *Policy environment: what kinds of policies support CBA mainstreaming?*

This research looked particularly at provisions for mainstreaming climate change adaptation within development and climate change policies, programmes and plans. The main aim was to identify the kind of policies, policy content and processes that support CBA mainstreaming. Past sectoral development policies (forest, agriculture and water resources), three-year interim development plans, climate change policy, and the National Adaptation Programme of Action (NAPA) and LAPA process and policies were analysed in terms of both content and process. These data were further supported by the outcomes of interviews and consultations with policy-makers, practitioners and communities.

The findings of this analysis suggest that climate change is not specifically addressed within sectoral plans and past development policies. Major development plans and sectoral policies in the agriculture, forestry and health sectors lack provisioning for climate change. Climate change seems not to have been prioritized in these sectors prior to 2007. However, the recent Three-Year Interim Plan (2010–2013) is more specific and comprehensive in terms of providing space to address climate change adaptation (Government of Nepal, 2010b), stressing the need for harmonizing sectoral plans by incorporating climate change issues. The specific provisioning for climate change adaptation in this interim plan creates opportunities for mainstreaming CBA. In terms of policy design, however, the majority of the development policies – including the Three-Year Interim Plan – were designed centrally, with limited consultation at both national and local levels.

The Ministry of Environment's climate change policy (2011a) has provisions for dealing with climate change issues. Although it does not specifically refer to mainstreaming climate change adaptation in development, it mentions linking and implementing climate adaptation with socio-economic development and income-generating activities to the greatest extent possible (Ministry of Environment, 2011a, p. 6). Adaptation is not given that much emphasis in the climate policy, however, despite the need for Nepal to prioritize this. Most policy-makers and participants consulted for this study were critical of the climate change policy. They felt it was very generic, outdated and driven by external agendas rather than country needs, because risk reduction was prioritized more than resilience building (interviews with policy-makers, January–March 2012). Existing literature also criticizes this climate change policy because of its failure to include key stakeholders in the process and its focus on approaches related to adjusting risk and 'climate proofing' against climate change (Helvitas, 2011).

The NAPA content and process, by contrast, looks favourable for scaling up and mainstreaming CBA. The document focuses on the interface between climate change adaptation and development and includes a section entitled 'National development planning as a framework for climate adaptation' (Ministry of Environment, 2010, p. 3). This discusses Nepal's development planning process and its responses to climate change issues. It also stresses specific links between the six thematic areas identified under the NAPA framework and national development goals (Ministry of Environment, 2010, p. 4). Interviews with policy-makers and practitioners revealed that these provisions within the NAPA reflect the importance the programme places on establishing links between adaptation

and development. Interviews also showed that ownership of the NAPA process amongst those consulted was very high. Previous research studies analysing the NAPA process in Nepal also suggest that Nepal's NAPA is highly participatory and inclusive (Ayers, 2011; Ghimire, 2011; Roberts, 2011).

The LAPA framework also looks very promising for supporting CBA in Nepal. This framework discusses the links between climate change adaptation and development and the impact of climate change on socio-economic development. The framework also recognizes that climate change vulnerability is specific to context, varying from place to place. The LAPA framework adopts bottom-up approaches and emphasizes the need for: communities to understand changing and uncertain future climatic conditions and engage effectively in the process of developing adaptation priorities; and for local adaptation priorities to be mainstreamed within local development planning and processes (Ministry of Environment, 2011b, Paragraph 4, p. 5). According to most of the policy-makers interviewed, this appreciation of the links to local planning processes could be a strategic entry point for mainstreaming CBA in development. It was also found that the LAPA design process was very localized and ensured community participation.

Analysis of key policies in terms of their focus suggests that one can find entry points within existing policies and systems for mainstreaming climate change adaptation. Challenges remain, such as the devolution of power and authority to local institutions, but building on successful practices, such as mobilizing local institutions and decentralization of decision-making, can benefit both climate change adaptation and development. Policies in sectors such as forestry and agriculture neglect climate change but look very promising in terms of addressing climate change risk and vulnerability, because they refer to practices, technologies and institutional mechanisms that can support successful implementation of adaptation priorities. There are also strong policy provisions for empowering community networks. Community forestry user groups and their network, mandated by the Forest Act of 1993, are locally popular and capable of managing the forest to which they are entitled (Acharya, 1999). Such experiences provide a strong basis for promoting CBA.

Similarly, the participants of the FGDs were asked to rate the statements derived by them. The ranking methods and indicators were discussed and identified by the policy-makers involved in the FGD. The focus group participants in this study felt that the entry point for CBA was based on the fact that development policies and plans tend to decentralize decision-making to local institutions and communities (see Table 1).

Discussions with policy-makers and practitioners revealed, however, that decentralization attempts in many other sectors in Nepal have been less successful. Policy

Table 1. Synergies amongst development and climate change policies in Nepal.

Major policies/plan	Focus on development (poverty alleviation)	Focus on climate change adaptation	Provision for decentralization
Three-Year Interim Plan	++++	+++	++++
Sectoral policies: forestry and agriculture	+++	None	++++
Climate change policy	++	++++	+++
NAPA	++++	+++++	+++
LAPA framework	+++	+++++	+++++

Source: FGD with policy-makers, February (2012).
Note: + is the lowest and +++++ is the highest ranked score provided by policy-makers.

implementation has failed to fully realize the ideals of self-governance, not only because of the vacuum of elected representatives in local government units, but also because of limits placed on the devolution of authority, funding and other resources (Gautam & Pokharel, 2011). Similarly, Dhungel (2011) found that decentralization policies have not changed the traditional way of operating at the local level. This kind of decentralization is called 'deconcentration', and

> merely involves the shifting of workload from central government ministry headquarters to staff located in offices outside of the national capital. Likewise it does not allow the local units ample freedom to take initiatives and decisions without the consent of the central authority. (Rondinelli et al., 1989, p. 76)

Those aiming to mainstream CBA must learn lessons from such past failures and devise ways of addressing them. One of the lessons from this for mainstreaming CBA is that decentralization alone is not enough, as it cannot guarantee the inclusion and empowerment of households and communities in the process. This mirrors the rich experiences and learning from the field of common property resource management, which stresses both decentralization and devolution as keys to the success of community-based natural resource management (Murphree, 2000; Reid, 2001). Devolution is relevant because it is a process whereby full authority and power is delegated to the local level, to empower vulnerable households and local communities and institutions to take decisions, and adaptation of practices that favour the inclusion of households and communities in the process (Litvack, Ahmad, & Bird, 1998, pp. 4–6). This implies that both decentralization and devolution are important policy reforms needed to mainstream CBA in development processes.

The local-level administration in Nepal can potentially handle its new role of managing climate change at the local level and use it to improve CBA. However, it depends upon the supportive role of the central government to fully delegate resources and provide capacity support to the local government to function effectively. Various scholars (Capistrano & Colfer, 2005; Ribot, 2006; Tacconi, 2007)

also confirm that decentralization shows promise in terms of increasing adaptive capacity at the local level and enhancing the role of local government and institutions to collectively deal with local issues.

3.2. Operationalizing mainstreaming: can existing mainstreaming approaches and mechanisms support scaling up of CBA in Nepal?

This section of the analysis looks at how climate change adaptation is mainstreamed in Nepal. The main objective of this section is to draw lessons on whether or not the existing mainstreaming approaches and mechanisms support scaling up CBA in Nepal. It specifically examines the contributions that two case studies – the PEI and the LAPA – make to the steps identified by Huq and Ayers (2008b) as critical for mainstreaming: increasing collaboration between agencies; increasing awareness of and knowledge on climate change at the local level; improving the knowledge base for informed decision-making; and integrating climate change adaptation into development planning.

3.2.1. Case study one: PEI

3.2.1.1. *Moving beyond silos: integrating and deepening collaboration.* The PEI shows that adopting a top-down process cannot successfully enhance local-level collaboration and synergy amongst stakeholders. The PEI programme document indicates that 'The key implementing partners for the PEI will be the National Planning Commission (NPC) and Ministry of Local Development' (Government of Nepal, 2010a, p. 15). This has limited the role of other stakeholders. Two practitioners interviewed felt that the PEI contributed little to bringing local stakeholders together and increasing collaboration between different agencies. They felt that the PEI was centralized, limited to a few agencies and overlooked the role of other stakeholders (interview with practitioners, January 2013).

The Dhading district case study showed that the issue of lack of collaboration amongst stakeholders had impacted on the implementation of policies and plans in benefit of local communities. For example, the PEI initiative intended to regulate the established – but unregulated – sand, gravel and stone industries along the riverbanks and reduce environmental destruction in the case study area, but failed to achieve progress because of a lack of support from local stakeholders. Communities consulted during the FGD in Jogimara VDC argued that they felt less motivated to participate in the activities that ignored their role. The majority of the practitioners interviewed also argued that in the absence of strong working collaborations, attempts to address complex issues like environmental and climate change are almost impossible.

The majority of participants in the two separate FGDs ($n = 30$), which involved representatives of local community groups in Jogimara VDC, argued that even with the implementation of climate change mainstreaming in local development planning, in practice, nothing has changed and the situation continues to deteriorate at the study site. The lack of collaboration amongst agencies had also impacted the response at the community level. For example, the local farmers' groups and cooperatives in Jogimara VDC said that they have not received any funding support from the government and other agencies in terms of implementing their community action plans.

3.2.1.2. *Information and knowledge: the basis for informed decision-making.* Information and knowledge on climate change in Nepal is very limited. Findings from interviews with policy-makers in this study show that the PEI carried out case studies on sand, gravel and stone and rural roads to investigate the impact of unplanned development interventions on the environment. These studies are also used to increase national and local awareness of the importance of mainstreaming poverty and environmental issues into development plans and programmes. The PEI also revised the Planning and Decision-Making Guidelines of the Local Bodies of Nepal, and the Minimum Condition and Performance Measures Guidelines (Poverty Environment Initiative, 2011).

The PEI did not, however, make any reference to the use of existing climate change information and databases when conducting various studies to inform policy-making and practice. The local government practitioners during the interview shared that, in the absence of information and knowledge on climate change, the integration of climate change adaptation practices in the annual development plan was difficult. They cited this as one of the reasons why mainstreaming climate change into local planning was challenging.

The lack of information and knowledge also had impacts at the community level whilst identifying local adaptation priorities. The majority of the communities

who participated in the FGDs and informal interactions ($n = 30$) revealed that they did not know what really caused the environment and climate change problems in the study area and how it could be addressed. They further argued that this is why the adaptation activities they included in their adaptation plan related more to their development experiences and everyday livelihood needs and may not specifically address climate-related risks and impacts. For example, the adaptation plan of one of the community groups in Jogimara VDC included activities such as control of open defecation, biodiversity conservation, conservation awareness and community-based seed bank management activities.

3.2.1.3. *Increased awareness, capacity and skills: the basics for mainstreaming.* PEI progress on awareness and capacity building at the local level is limited. The PEI annual progress report mentions that,

> … orientation trainings and resource materials provided to community organizations in 2011 have empowered these groups to demand the integration of poverty-environment climate concerns in local planning and budgeting processes. (Poverty Environment Initiative, 2012, p. 3)

The report cites examples, such as

> … in response to the demand of Ward Citizen Forum, one small community-based organization, whose members include women and representatives from marginalized communities, and community forest user groups, the Jogimara VDC, Dhading has for the first time approved climate change adaptation activities to be implemented in 2012. (p. 3)

Data from the field observation and FGDs with the communities and local officials in Jogimara VDC also showed that besides training, no further activities had been conducted. Communities in Jogimara VDC ($n = 30$) revealed that none of them were involved in orientation and capacity building activities conducted by the local government on climate change. The agriculture group who prepared their climate change adaptation activities received support from a NGO but had no support from the local government. The majority of the FGD participants argued that most of the awareness and capacity building activities in fact do not reach actual households, but often focus on a few handpicked individuals and organizations.

According to the majority of the communities and practitioners consulted, the capacity building approach of the PEI case had little impact because it consisted of limited activities that focused on only a few individuals within the government system. Some practitioners stated that, in the absence of the required skills and capacity, they were

21

reluctant to include climate change adaptation activities within their local development plan.

3.2.1.4. *Institutionalization and scaling up: sustaining mainstreaming.*

Project documentation shows that the PEI seems to have strategic influence over government planning processes. The National Planning Commission has adopted a climate resilience framework and integrated addressing climate change into the Three-Year Interim Plan. Similarly, the Ministry of Local Development addressed environment and climate change in its monitoring framework and appraisal system. Recently, the Ministry of Finance developed a budgetary code for climate change expenditure. These policy changes resulted from PEI interventions at the national level. However, there were very few attempts by other ministries and stakeholders to mainstream climate change into their annual planning processes, because these agencies were not engaged actively.

One of the issues that emerged from local-level discussions relates to the approach to mainstreaming. Currently, PEI takes top-down planning as the entry point to mainstreaming climate change adaptation measures in development. Most practitioners and communities interviewed and consulted said that current government planning and service delivery structures are constrained in terms of effectively responding to local needs as well as increasing the direct access of communities and households to services and resources. Communities consulted during the FGD of Jogimara VDC said that the annual development planning cycle is rigid and accountability is upwards and towards government agencies only, as opposed to downwards towards vulnerable communities. They further argued that long-term climate change issues and associated uncertainties cannot be dealt with because the annual planning cycle is limited in terms of time and resources. They suggested reforming the planning process to suit the climate change context.

There was evidence at the local level where the current government structure and planning acted as a barrier to scale up community initiatives. For example, during informal interactions and the first FGD with communities in Jogimara VDC, it was revealed that the agriculture group of the VDC, which had prepared a local-level climate change adaptation plan, could not implement its adaptation priorities because of a lack of support from the local government. Due to the fixed planning cycle and rigid structure, the local VDC officials could not accommodate and integrate the communities' adaptation plans within the annual development planning process. According to the communities, the VDC even failed to provide technical and financial resources to implement the adaptation plan of the group. The findings imply that in the absence of local ownership and government support, mainstreaming CBA remains challenging.

3.2.2. *Case study two: LAPA*

3.2.2.1. *Moving beyond silos: integrating and deepening collaboration.*

The LAPA case study shows that the process has helped to raise awareness amongst local institutions and brought grassroots institutions together. The major stakeholders involved in the LAPA included: government, communities, NGOs, political parties and other social organizations. Amongst the stakeholders, communities' representation was relatively high in the LAPA preparation. Most FGD participants in Sukrauli, Bangesaal and Dhungegadi VDCs ($n = 120$) at the local level felt they had contributed to the adaptation planning process. Similarly, more than 64% of practitioner respondents ($n = 18$) felt that the LAPA had been instrumental in bringing different agencies together.

Field observations showed that many community-based organizations and households were part of the LAPA design and piloting process. The majority of the practitioners confirmed that at least 20–25% of the total households were directly engaged in the adaptation design process. Communities were observed assessing climate change issues, identifying adaptation priorities and suggesting institutional and financial mechanisms for implementing adaptation interventions. Most of the participants of the FGD in all three VDCs argued that the level of collaboration between community groups and local government agencies and NGOs has increased due to implementation of LAPA.

The LAPA pilot site showed positive outcomes from increased collaboration amongst stakeholders. This research found specific examples of resource leveraging as an outcome of improved collaboration between agencies. In the case of Terai District, a basket fund was established by each Village Forest Coordination Committee (VFCC). According to the communities and practitioners consulted in Nawalparasi district, a total of 35,097,490 (equivalent to 449,967 USD) Nepali Rupees was channelled from the basket funds to the community level prior to January 2012 from this fund. The participants of FGDs in Dhungegadi VDC ($n = 40$) revealed that the local communities were active in demanding services and communicating their needs and priorities to the service providers. Table 2 shows that almost all the community groups in Dhungegadi VDCs were successful in drawing down financial resources from local government and other agencies to implement their adaptation priorities.

Whilst the LAPA case study shows that collaboration amongst agencies is very high at the community and VDC level, it becomes more complicated at the district level and above. Cooperative work is still lacking in government institutions, particularly at the district level. District-level stakeholders struggled to find the right strategies and action plans to deal with climate change. Discussions revealed that this was in part due to the lack of

Table 2. Examples of community groups in Dhungegadi VDC receiving financial support to implement some of their adaptation priorities.

Types of activity funded	Groups receiving the fund	Amount (NPR) total 584,990 = 7780 USD
Bacchidanda water tank construction and emergency fund	Kalidhunga ward number 9	73,894
Khaltak Pani water tank construction	Ward number 5	50,886
Revolving fund and emergency fund	Khahare community forest	67,000
Revolving fund for supporting adaptation	Kamaladi ward number 6	60,000
Revolving fund for supporting adaptation activities	Bachimpokhara	35,000
Provision of fire control material	Kamaladevi, ward number 6	7500
Water tank construction	Bagedi, ward no 3	70,000
Gatkhola water tank construction	Chaukedawla, ward number 7	70,000
Dharadikhola irrigation support	Chabise, ward number 4	70,000
Plastic pond for water collection	Rangbang, ward number 8	51,210
Stretcher and medicine for emergency	Jaspur, ward number 8	29,500
Emergency fund establishment	VDC level	25,000

Source: Author.

central government guidelines and also ambiguity regarding what decisions to make due to a lack of information and knowledge on climate change (District-level multi-stakeholder brainstorming workshop, 13th January 2012). In addition, the bottom-up approaches adopted by the LAPA were limited because they focused more on NGOs and communities, and less on government. This is a strong reminder of the importance of involving a diversity of actors to address climate change issues, ranging from local communities to policy-makers.

3.2.2.2. *Information and knowledge: the basis for informed decision-making.* The findings show that the initial 18 months of the LAPA piloting process focused on identifying opportunities and challenges for mainstreaming. Several background studies were carried out to increase understanding of the local governance system, institutional and financial mechanisms, successful practices and available tools and approaches. According to the majority of the community groups consulted ($n = 6$), the information generated and the testing of participatory approaches and tools provided a strong basis for designing and piloting the LAPA framework. One government official also argued that the systematic studies helped convince government policy-makers, practitioners and communities about the importance of supporting locally driven climate change adaptation (interview with policy-maker).

In the case of the LAPA, the review of adaptation planning processes and the plan itself indicated that adaptation planning and prioritization of adaptation options were based on the available information and knowledge of communities and staff from NGOs, and government organizations. Communities and practitioners relied on their experience of climate variability and current risk factors to devise adaptation options (FGD with practitioners in Pyuthan district). Whilst experience of development interventions is very valuable in terms of identifying adaptation options, the scale and magnitude of impacts expected

demand scientific knowledge and innovative technologies and practices to deal with climate extremes. The discussions with practitioners and communities in the two districts suggest that this sharing of knowledge and learning is lacking in the LAPA process.

The lack of information and knowledge was having an impact on the identification and prioritization of adaptation activities at the community level. The participants in all four FGDs carried out in Dhungegadi and Bangesaal VDCs revealed that investment was made mostly in infrastructure-related activities. In Dhungegadi alone, 60.5% of the funding support was invested (out of 876,690 NPR = 11,690 USD) in construction and maintenance of drinking water sources, a pipeline and reserve tank. Whereas in Bangesaal VDC, more than 67% (1,360,956 NPR = 18,146 USD) was invested in drinking water facilities and maintenance of irrigation channels. Communities argued that in the absence of knowledge and information about the technology and practices suitable for dealing with climate change activities, they preferred general development priorities.

The study data showed that the investment made in general development activities had only limited household and geographical coverage. The communities also revealed that the general development activities may not directly benefit the poor and vulnerable households:

> There is now dissatisfaction among us regarding the allocation of resources and selection of investment areas. The resource allocation was made considering the group request rather than assessing the actual risk and impact of climate change. The community forestry groups that were vocal and influential received the support. Within the group also, the poor households in fact had nothing in their plate. (FGD participant in Dhungegadi VDC)

Analysis of the LAPA case study reveals that not enough has been done in terms of climate change information and knowledge generation. The LAPA has made comparatively good progress in terms of developing participatory methodologies and approaches to fill knowledge gaps. However, local decisions on climate change relied on existing information on climate variability. To address the scale and magnitude of future climate risks, additional effort will be required, involving stronger interaction between households, communities and the scientific/research community.

3.2.2.3. *Increased awareness, capacity and skills: the basics for mainstreaming.*

The findings show that LAPA initiatives have mobilized large numbers of communities, government line agencies, NGOs, donors and community-based organizations. Field observations and discussions with communities and practitioners in all three VDCs revealed that the LAPA process has enriched the interest of communities and community-based institutions at the local level. From the field observations, it was noted that local and CBA planning has been instrumental in sensitizing communities and households on climate change issues. Secondary information derived from project documentation indicates that a total of 9 events were organized in Pyuthan district, which sensitized 610 women and 560 men. Of these, 595 people were from disadvantaged groups, 450 from ethnic minorities and 50 from religious minorities.

The evidence gathered from community FGDs shows that the local and CBA planning was instrumental in sensitizing communities and households on issues related to climate change. It was found that the level of interest and awareness of local stakeholders had increased due to capacity building activities, and a large number of community groups in the research areas used the information and knowledge gained from the training to deal with climate change issues more systematically. For example, 31 community groups in Bangesaal, Dhungegadi and Sukrauli VDCs had prepared local and community adaptation plans (CAP) based on the training guidance to address climate change issues. These community groups had also formed an institutional mechanism, for example a VFCC, to coordinate the adaptation action carried out by different community-based institutions (see Table 3).

Field observation and interaction with communities and practitioners in Pyuthan district revealed that the LAPA piloting had also attracted the interest of communities and community-based institutions at the local level on climate change adaptation. More than 90% of community

respondents in the FGD in Bangesaal VDC and 92% of respondents in Dhungegadi VDC were satisfied with their exposure on climate change issues and their involvement in climate change adaptation planning as a result of the awareness activities. They further stated that due to local-level awareness on climate change issues, it has been easier for their community groups to receive support from external agencies such as NGOs and government. Many respondents, however, raised the need for better skills transfer and sustained capacity building.

Despite some good progress on raising awareness at the local level by the LAPA pilot activities, capacity and knowledge transfer remained challenging. This research found that knowledge and skills transfer to service providers, particularly government agencies, was limited. For example, the NGOs who were involved in facilitating local-level adaptation planning do possess experience and knowledge of the subject matter, but did not transfer this to the government service providers (mostly at the district level). Most of the practitioners ($n = 24$) argued that there is a lack of technical and human resources at the district level to help communities to respond well to climate change.

Similarly, there were issues relating to the identification of technologies and practices that best suit community needs. Communities and practitioners consulted felt that identification of adaptation options was ad hoc and constrained by limited knowledge and information about future climate risks and impacts. The majority of the practitioners interviewed (90%) revealed that they were reluctant to offer support to communities because of their limited capacity and skills on climate change adaptation. The findings imply that limited skill and capacity at the local level has institutional and policy implications for the scaling up of CBA within government institutions.

3.2.2.4. *Institutionalization and scaling up: sustaining mainstreaming.*

Findings from the interviews with communities in this study show that strong local institutional ownership was ensured during local mainstreaming activities. Community forestry user groups were mobilized as grassroots community groups to coordinate adaptation initiatives at the household and community level. Similarly, at the VDC level, VFCCs were formed under the chairmanship of the VDC secretary, involving all forest user-group

Table 3. Number of community groups involved in adaptation activities.

Research site	Number of community groups involved in adaptation	Number of households benefited	Vulnerable households involved	Number of coordination mechanisms established
Bangesaal	11	1521	827	1
Dhungegadi	9	1177	579	1
Sukrauli	11	335	250	1

Source: Author.

Table 4. Scaling up the LAPA in Nepal.

Organization involved in scaling up LAPA	Number of LAPAs formed/targeted
Interim Forestry Programme/UK Department for International Development and Swiss Development Cooperation	298 LAPAs and 1468 CAP
World Wildlife Fund	3 LAPAs
Nepal Climate Change Support Programme	70 VDC-level LAPAs

Source: Author.

representatives, political parties, government service providers and local NGOs. Most of the communities in the FGD said that climate change adaptation is already part of their community-level plans and they are committed to implement the priorities identified in the plans.

The district-level FGD with practitioners and communities in Pyuthan district showed that the learning of Bangesaal and Dhungegadi had been scaled up in other areas of the district. A total of 58 CAP and 3 LAPAs had been prepared in the whole district up until 2012. Eleven VFCCs had been formed to support climate change adaptation work at the local level. In addition, the FGD with practitioners in Nawalparasi district indicated that learning of LAPA experiences in Sukrauli VDC has now been scaled up to other areas. A total of 88 VDC-level LAPAs and 165 community-level adaptation plans had been prepared and implemented in the district by 2012.

The available literature and field observations suggest that LAPA piloting has also been scaled out to other areas of Nepal and some bilateral donors are taking it forward. Table 4 demonstrates that LAPA piloting – particularly at the research sites – has geographically reached more than 40% of the country and 10% of VDCs.

There have been several challenges, however, related to scaling up CBA through the LAPAs. Due to resource constraints, most plans prepared at the community level have not been implemented. Communities also raised issues about the lack of funding and its impact on the implementation of priorities identified in their adaptation plans. According to the participants of FGD in Sukruali VDC, due to financial constraints, 90% of their identified adaptation priorities have not been implemented so far.

The financial analysis of two VDC-level LAPAs in Bangesaal and Dhungegadi (Pyuthan District) shows that adaptation funding requirements are great but the current support is minimal. The Dhungegadi and Bangesaal LAPAs project requirements are US$1.5 and US$2.4 million, respectively, for the next 5 years to implement only the most urgent and immediate adaptation priorities. The funding available for implementing the LAPA until 2012, however, was only US$6772 for Dhungegadi and US$5591 for Bangesaal. Donors and government seem to focus on preparing plans and moving into new areas without ensuring sustainability of the initiatives already started. This presents the challenge of how to ensure mainstreaming is sustainable and supports the needs of vulnerable communities and households.

The findings show that whilst there have been some initial efforts to mainstream climate change adaptation into development in Nepal, progress is limited.

Conclusion

This study looked into the policy environment and practices of mainstreaming in Nepal in order to identify operational mechanisms for scaling up CBA to Climate Change. The evidence shows that although Nepal's major sectoral policies lack a clear focus on CBA, the decentralization of power and authority to local institutions has progressed significantly over the last few years with regard to the sectoral policies. The success stories of decentralization and community development provide a favourable policy environment for mainstreaming CBA in development. However, it was also found that most of the policies were not being translated into action and were not in favour of the poor and vulnerable households.

The findings also reveal that there were both opportunities for and constraints to mainstreaming CBA in development in Nepal. The two mainstreaming initiatives analysed for this research demonstrate differences in mainstreaming progress due to the approach and strategy applied by each. The PEI has progressed more with respect to policy influence and less in terms of putting mainstreaming into action. Influencing policy and changing existing environmental regulation has occurred at the central level but policies have not been translated into action. In contrast, the LAPA mainstreaming initiative was successful in mobilizing local community groups and increasing their awareness. However, mainstreaming CBA in development at the operational level was constrained by: limited scientific information on climate change; limited knowledge on technology and practices of adaptation; limited awareness amongst practitioners and communities; a rigid planning structure; and limited financial resources.

The evidence suggests that there is a need for both integrated and inclusive policies in order to ensure wider participation in policy-making as well as efficient institutional, financial and knowledge mechanisms to effectively mainstream CBA in development. The most suitable operational mechanism specifically for Nepal, as argued in this research, is to have: (a) an overarching, integrated and locally accountable climate change and development policy that can achieve both development and climate change goals; (b) integrated and multi-stakeholder owned governance and institutional and financial mechanisms at the national and district levels; and (c) locally inclusive and responsive institutional and delivery structures that address the equity issues and varied adaptation demands of households and communities.

References

Acharya, K. (1999). Twenty four years of community forestry in Nepal. *International Forestry Review, 4,* 149–156. doi:10.1505/IFOR.4.2.149.17447

Agrawal, A. (2008). *The role of local institutions in adaptation to climate change.* Working Paper (W08I-3). Washington, DC: International Forestry Research and Institutions Program (IFRI), pp. 1–63.

Agrawala, S. (2004). Adaptation, development assistance and planning: Challenges and opportunities. *IDS Bulletin, 35*(3), 50–54. doi:10.1111/j.1759-5436.2004.tb00134.x

Agrawala, S. (2005). *Bridge over troubled waters: Linking climate change and development.* Paris: Organization for Economic Cooperation and Development, pp. 1–18.

Ayers, J. (2011). *Understanding the adaptation paradox: Can global climate change adaptation policy be locally inclusive?* (PhD). London: London School of Economics and Political Science.

Ayers, J., Alam, M., & Huq, S. (2010). Global adaptation governance beyond 2012. Developing country perspectives. In F. Biermaan, P. Pattberg & F. Zelli (Eds.), *Global climate governance beyond 2012: Architecture, agency and adaptation* (pp. 270–285). Cambridge: Cambridge University Press.

Ayers, J., & Huq, S. (2009). Supporting adaptation to climate change: What role for official development assistance? *Development Policy Review, 27*(6), 675–692. doi:10.1111/j.1467-7679.2009.00465.x

Ayers, J., & Huq, S. (2013). Adaptation, development and the community. In J. Palutikof, S.L. Boulter, M.S.S. Andrew, J. Ash, M. Parry, & D.G. Marie Waschka (Eds.), *Climate adaptation futures* (pp. 203–214). Oxford: John Wiley and Sons Ltd Sussex.

Ayers, J.M., Huq, S., Faisal, A.M., & Hussain, S.T. (2013). Mainstreaming climate change adaptation into development: A case study of Bangladesh. *Wiley Interdisciplinary Reviews: Climate Change, 5,* 37–51. doi:10.1002/wcc.226

Capistrano, D., & Colfer, C.J.P. (2005). *Decentralization: Issues, lessons and reflections. The politics of decentralization: Forests, power and people.* London: Earthscan.

Dhungel, D. (2011). *Decentralization in Nepal: Laws and practices* (NIBR-rapport 2011: 23). Norway: NIBR.

Gautam, M., & Pokharel, B. (2011). *Foreign aid and public policy process in Nepal: A case of forestry and local governance.* Kathmandu: South Asia Institute of Advanced Studies.

Ghimire, S. (2011). *Climate justice: Bottlenecks and opportunities for policy-making in Nepal.* Kathmandu: South Asia Institute of Advanced Studies.

Government of Nepal. (2010a). *Programme document – Poverty Environment Initiatives (PEI).* Kathmandu: GON/UNDP/UNEP.

Government of Nepal. (2010b). *Three year interim plan.* Kathmandu: National Planning Commission.

Helvitas. (2011). *Nepal's climate change policies and plans: Local communities perspective.* Kathmandu: Swiss Intercorporation Nepal.

Huq, S., & Ayers, J. (2008a). *Mainstreaming adaptation to climate change in least developed countries (LDCs).* London: IIED.

Huq, S., & Ayers, J. (2008b). *Taking steps: Mainstreaming national adaptation* (IIED Briefing). London: IIED.

Huq, S., & Reid, H. (2007). *Community based adaptation* (IIED Briefing). London: IIED.

Huq, S., Reid, H., Konate, M., Rahman, A., Sokona, Y., & Crick, F. (2004). Mainstreaming adaptation to climate change in least developed countries (LDCs). *Climate Policy, 4*(1), 25–43. doi:10.1080/14693062.2004.9685508

Huq, S., Reid, H., & Murray, L.A. (2006). *Climate change and development links.* London: Sustainable Agriculture and Rural Livelihoods Programme, IIED.

Lasco, R.D., Pulhin, F.B., Patricia, A.S., Delfino, R.P., Roberta, G., & Garcia, K. (2009). Mainstreaming adaptation in developing countries: The case of the Philippines. *Climate and Development, 1*(2), 130. doi:10.3763/cdev.2009.0009

Lebel, L., Li, L., Krittasudthacheewa, C., Juntopas, M., Vijitpan, T., Uchiyama, T., & Krawanchid, D. (2012). *Mainstreaming climate change adaptation into development planning.* Bangkok: Adaptation Knowledge Platform and Stockholm Environment Institute.

Litvack, J.I., Ahmad, J., & Bird, R.H. (1998). *Rethinking decentralization in developing countries.* Washington, DC: World Bank Publications.

Lobo, C. (2011). *Mainstreaming climate change adaptation: The need and role of civil society organization.* Pune: Watershed Organization Trust (WOTR).

Ministry of Environment. (2010). *National Adaptation Programme of Action (NAPA).* Kathmandu: Nepal Ministry of Environment.

Ministry of Environment. (2011a). *Climate change policy, 2011.* Kathmandu: Government of Nepal.

Ministry of Environment. (2011b). *Local Adaptation Plan of Action (LAPA) framework.* Kathmandu: Ministry of Environment.

Murphree, M.W. (2000). Communal approaches to natural resource management in Africa: From whence and to where? *Journal of International Law and Wildlife Policies, 7*(3–4), 203–216. doi:10.1080/13880290490883250

Oliver, P., & Jupp, V. (2006). Purposive sampling. In V. Jupp (Ed.), *The SAGE dictionary of social research methods* (pp. 244–245). New Delhi: Sage Publications.

Persson, Å. (2008). *Mainstreaming climate change adaptation into official development assistance: A case of international policy integration.* EPIGOV Papers No. 36. Berlin: Ecologic – Institute for International and European Environmental Policy, pp. 1–31. Retrieved from http://www.sei-international.org/mediamanager/documents/Publications/Policy-institutions/epigov_paper_36_persson.pdf

Poverty Environment Initiative. (2011). *Progress report of Poverty Environment Initiatives (PEI).* Kathmandu: Ministry of Local Development.

Poverty Environment Initiative. (2012). *Progress report of Poverty Environment Initiatives (PEI).* Kathmandu: Ministry of Local Development.

Reid, H. (2001). Contractual national parks and the Makuleke community. *Human Ecology, 29*(2), 135–155. doi:10.1023/A:1011072213331

Reid, H., Huq, S., & Murray, L. (2010). *Community champions: Adapting to climate challenges.* London: IIED.

Reid, H., & Schipper, E.L.F. (2014). Upscaling community-based adaptation: An introduction to the edited volume. In E. Lisa, F. Schipper, J. Ayers, H. Redi, S. Huq, & A. Rahman (Eds.),

Community based adaptation to climate change: Scaling it up (pp. 3–21). New York: Routledge.

Ribot, J.C. (2006). Choose democracy: Environmentalists socio-political responsibility. *Global Environmental Change, 16* (2), 115–119. doi:10.1016/j.gloenvcha.2006.01.004

Ridder, D., & Team, H. (2006). *Learning together to manage together: Improving participation in water management.* Osnabrueck, Germany: University of Fachbereich Mathematik.

Roberts, E. (2011). *Addressing vulnerability to climate change in least developed countries? An evaluation of the national adaptation programmes of Bangladesh, Nepal, Malawi and Tanzania* (MSc). London: London School of Economics.

Rondinelli, D.A., McCullough, J.S., & Ronald, J.W. (1989). Analysing decentralization policies in developing countries: A political-economy framework. *Development and Change, 20*(1), 57–87. doi:10.1111/j.1467-7660.1989.tb00340.x

Sjoberg, G., Williams, N., Vaughan, T.R., & Sjoberg, A.F. (1991). The case study approach in social research. In J.R. Feag, A.M. Orum, & G. Sjoberg (Eds.), *A case for case study* (pp. 27–79). North Carolina: The University of North Carolina Press.

Srinivasan, A., & Uchida, T. (2008). Mainstreaming and financing adaptation to climate change. In A. Srinivasan (Ed.), *The climate regime beyond 2012: Reconciling Asian development priorities and global climate interests* (pp. 57–83). Hayama, Japan: Institute for Global Environmental Strategies.

Tacconi, L. (2007). Decentralization, forests and livelihoods: Theory and narrative. *Global Environmental Change, 17*(3), 338–348. doi:10.1016/j.gloenvcha.2007.01.002

Walt, G., & Gilson, L. (1994). Reforming the health sector in developing countries: The central role of policy analysis. *Health Policy and Planning, 9*(4), 353–370. doi:10.1093/heapol/9.4.353

REVIEW ARTICLE

Farmers, food and climate change: ensuring community-based adaptation is mainstreamed into agricultural programmes

Helena Wright[a], Sonja Vermeulen[b], Gernot Laganda[c], Max Olupot[d], Edidah Ampaire[e] and M.L. Jat[f]

[a]Centre for Environmental Policy, Imperial College London, London, UK; [b]CGIAR Research Program on Climate Change, Agriculture and Food Security (CCAFS), Coordinating Unit, University of Copenhagen, Copenhagen, Denmark; [c]International Fund for Agricultural Development, Rome, Italy; [d]African Forum for Agricultural Advisory Services, Kampala, Uganda; [e]International Institute of Tropical Agriculture (IITA), Kampala, Uganda; [f]International Maize and Wheat Improvement Centre (CIMMYT), New Delhi, India

Climate change creates widespread risks for food production. As climate impacts are often locally specific, it is imperative that large-scale initiatives to support smallholder farmers consider local priorities and integrate lessons from successful autonomous adaptation efforts. This article explores how large-scale programmes for smallholder adaptation to climate change might link effectively with community-led adaptation initiatives. Drawing on experiences in Bangladesh, Mozambique, Uganda and India, this article identifies key success factors and barriers for considering local priorities, capacities and lessons in large-scale adaptation programmes. It highlights the key roles of extension services and farmers' organizations as mechanisms for linking between national-level and community-level adaptation, and a range of other success factors which include participative and locally driven vulnerability assessments, tailoring of adaptation technologies to local contexts, mapping local institutions and working in partnership across institutions. Barriers include weak governance, gaps in the regulatory and policy environment, high opportunity costs, low literacy and underdeveloped markets. The article concludes that mainstreaming climate adaptation into large-scale agricultural initiatives requires not only integration of lessons from community-based adaptation, but also the building of inclusive governance to ensure smallholders can engage with those policies and processes affecting their vulnerability.

1. Introduction

Climate-related risks and opportunities play a prominent role in agricultural development, but are not always recognized in sector programming and investment planning. Climate change is now affecting crop productivity and the ability of farmers to harvest and process agricultural produce, with direct impacts on the nearly 70% of people in developing countries living in rural areas where agriculture is the main livelihood (Vermeulen, Campbell, & Ingram, 2012). In many rural areas, episodes of extreme weather interrupt access to markets, while restricted livelihood options and insufficiently diversified energy systems perpetuate the degradation of those ecosystems which are needed more than ever as natural buffers against floods, landslides and soil erosion. Climate-related disasters can disrupt social networks and wipe out years of financial savings, rolling back decades of development progress (Carter, Little, Mogues, & Negatu, 2007). Climate change materializes predominantly as a threat multiplier for poor rural households, adding new dimensions to the portfolio of risks, opportunities and longer term trends facing people whose livelihoods depend on agriculture.

Adaptation responses to these risks and trends have been distinguished in the literature between 'autonomous adaptation' at the individual, household or farm level and 'planned adaptation' usually at the level of national government. In reality these levels are functionally linked (Adger, Huq, Brown, Conway, & Hulme, 2003; Eriksen et al., 2011), but nonetheless the distinction between them provides a useful framework for distinguishing two contrasting mechanisms for mainstreaming community-based adaptation (CBA): either by 'bottom-up' scaling up or out from local community-driven, autonomous adaptations, or alternatively by being built into the design of 'top-down' planned adaptation programmes. Both of these mechanisms offer opportunities to integrate community-based approaches to climate change adaptation with larger agricultural planning and investment processes, thereby benefitting a high number of farmers at large geographic scales. As the impacts of climate change are felt locally, there is a growing body of research arguing that community-identified and -led activities are integral to effective adaptation for smallholder farmers (Heltberg, Siegel, & Jorgensen, 2009; Kansiime, 2012; Reid et al., 2009). However, as we elaborate below, the transfer of knowledge and practice from local adaptation experience appears to be rare, despite its potential value.

At present it is unclear to what extent CBA has been mainstreamed into agricultural programmes and funding agencies across different levels. The purpose of this article is therefore to use insights from four case studies to explore the barriers to and opportunities for mainstreaming CBA in agriculture at broader scales. The next section defines and explores the implications of key terms such as CBA, mainstreaming and scaling up, as well as summarizing progress at national and international levels. Section 3 summarizes four cases: two experiences of mainstreaming adaptation into agricultural planning in Mozambique and Bangladesh, and two local-level examples from Uganda and India in which community-based agricultural adaptation endeavours have made efforts to influence policy at higher levels (to scale up). We analyse the opportunities and barriers for community-based approaches to provide effective inputs to higher level policy or larger scale programmes, and for large-scale programmes to respond effectively. Section 4 considers the key role of extension as a conduit for two-way learning between the national and community levels, and concludes with some emerging suggestions for more effective exchange and co-learning.

2. Mainstreamed versus community-based adaptation: from gaps to meeting points

Climate change mainstreaming is now widely promoted as a more effective approach than stand-alone interventions on climate change adaptation or mitigation. In its broadest sense, climate change mainstreaming entails incorporation of climate change considerations into public policy and practice, at all planning levels, across all sectors and involving public, private and civil society actors. Substantial practical guidance now exists for policy-makers and practitioners to mainstream climate adaptation into development policies and programmes, including specifically in the agriculture sector (CARE, 2009; FAO, 2012a; UNDP-UNEP, 2011).

Most of this mainstreaming guidance emphasizes the importance of engaging stakeholders from the start, including local stakeholders in affected communities, and being responsive to their expressed priorities and needs. Yet most guidance fails to clarify whether the mainstreaming process is envisaged as a bottom-up process through which successful CBA actions are replicated and scaled-up to reach new target groups and geographic areas, or as a top-down process in which large-scale government

action plans deliver pre-determined adaptation benefits. This question is especially relevant as it helps to determine the ambition and institutional layout of a climate mainstreaming process: such a process could entail the comparatively straightforward dissemination of new technical information and know-how through established channels (such as extension services), or the more complex strengthening of innovative processes and institutional mechanisms through which relevant adaptation knowledge can be generated or internalized.

In principle, agricultural programmes that mainstream adaptation into their planning do not preclude community-based approaches, but there may be considerable challenges in reaching scale while also assuring local 'ownership' (control over decisions and resources) and accommodating the diversity that comes with differing local priorities. CBA can be defined as 'a community-led process, based on communities' priorities, needs, knowledge, and capacities, which should empower people to plan for and cope with the impacts of climate change' (Reid et al., 2009, p. 13). CBA builds on a long history of community-driven approaches to development, which rose to global prominence in the 1990s, following perceptions that large-scale centralized development programmes were performing poorly, and that poor people could and should be the central decision-makers in their own development (Mansuri & Rao, 2004). It has been suggested using farmers as the agents of change is more likely to be sustainable, while recognizing pitfalls relating to maladaptation, that is, potential negative impacts of adaptation across spatial or temporal scales (Vincent et al., 2013). While policies proliferate at the national level, which can lead to duplication, households tend to adapt to multiple stresses in an integrated way (Stringer, Mkwambisi, Dougill, & Dyer, 2010). Community-driven development, in which decisions are made and budgets are allocated locally, can deliver sustainable outcomes at lower cost than centrally managed programmes, while also conferring better governance in terms of local accountability, transparency and empowerment (Binswanger-Mkhize, De Regt, & Spector, 2010). Yet participation does not automatically guarantee success, which depends additionally on sustained facilitation and careful design (Blaikie, 2007; Mansuri & Rao, 2004) as well as an enabling policy environment.

Most CBA endeavours do not arise from large-scale programmes but rather from local innovation, either driven purely by communities or else prompted and facilitated by a non-governmental organization (NGO), development agency or research organization operating at the local scale. For these endeavours to reach scale presents a different set of challenges than for large-scale mainstreamed programmes to be responsive to community-level priorities. Some barriers identified from global experiences in scaling up community-driven development initiatives included hostile institutional settings, barriers to accessing finance, lack of compatible incentives, stakeholders with differing values, geographical or socio-political differences and logistical challenges (Binswanger-Mkhize et al., 2010). However, it must not be assumed participative processes are generating an effective and consensual output: the term 'participation' is often used to describe 'very rudimentary levels of consultation between professionals and community members' (Taylor, 2003, p. 122), in contrast to active engagement involving two-way information flows (Reed et al., 2009).

Arguably, a supportive enabling environment with participation of local communities is required to integrate adaptation into development (Sietz, Boschutzz, & Klein, 2011). Scaling up means more than just *physical scaling up* (mass replication); but also *social scaling up* (increasing social inclusiveness) and *conceptual scaling up* in terms of moving beyond participation to embedding empowerment in the entire development process (Binswanger-Mkhize et al., 2010). This links to the concept of 'procedural justice' in adaptation (Thomas & Twyman, 2005), leading us to seek not only mainstreaming of climate change adaptation into agriculture, but also inclusive governance whereby farming communities can engage with policies and processes affecting their vulnerability. Adapting our conceptual framework from Linn (2012), we argue mainstreaming CBA requires an 'enabling environment' with institutional, political, fiscal, market, resource, cultural and learning space for CBA to occur as a process. Mainstreaming CBA is more than scaling up of specific adaptation practices or knowledge, it is about mainstreaming institutional and organizational approaches that allow this knowledge to be generated.

In policy and practice, there has been some progress in mainstreaming adaptation in agriculture. Under the United Nations Framework Convention on Climate Change (UNFCCC), many least developed countries identified agriculture as a vulnerable sector in their National Adaptation Programmes of Actions (NAPAs). Countries are now beginning to mainstream adaptation into agricultural policies and programmes through National Adaptation Plans (NAPs), which should enable the adoption of a 'participatory and fully transparent approach, taking into consideration vulnerable groups, communities and ecosystems' (UNFCCC, 2011, p. 80). Multi-lateral funds and financial agencies are in the early stages of integrating mechanisms for managing climate risks and trends into their programming. The World Bank has guidelines for mainstreaming of adaptation into agriculture and natural resources management projects (World Bank, 2010), while the International Fund for Agricultural Development (IFAD) recently established the world's single largest fund for adaptation in smallholder agriculture. IFAD's new Adaptation for Smallholder Agriculture Programme (ASAP) provides a new source of grant co-financing to scale up and integrate adaptation across IFAD's

approximately US$1 billion annual new investments and introduces a systematic appraisal of climate-related risks and vulnerabilities into agricultural investment planning (IFAD, 2012a). ASAP investments under development at the country level include a range of CBA components, including participatory mapping and vulnerability assessment, delegation of priority-setting for spending of adaptation funds to community groups, and mechanisms for community-to-community learning across administrative and geographic boundaries.

3. Mainstreaming CBA in agriculture: insights from case studies

Case studies were selected that cover a range of different geographical regions and climate change vulnerabilities across four countries in Asia and Africa. Experiences are drawn from national-level experiences in mainstreaming adaptation into agricultural planning in Mozambique and Bangladesh using secondary sources, and from local (sub-national) level experiences in Uganda and India using primary data. Each case study introduces the relevant institutions, mechanisms and project activities (whether governmental or provided by NGOs), drawing lessons from these in terms of the success factors and barriers encountered in mainstreaming CBA into agricultural programming, and exploring the policy implications.

3.1. *Mainstreaming of CBA in agriculture in Bangladesh*

In Bangladesh, initial steps have been taken to mainstream processes and lessons from CBA, but there are various institutional and communication barriers. Experiences with mainstreaming climate change adaptation into agricultural planning under the 'Livelihood Adaptation to Climate Change' (LACC) project were reviewed using secondary literature and reports. Located in the low-lying Ganges–Brahmaputra delta, Bangladesh is at risk of increasing flooding, more intense cyclones and sea level rise in a warmer climate (Huq, Rahman, Konate, Sokona, & Reid, 2003). The LACC project under the Comprehensive Disaster Management Programme (CDMP) promoted livelihood adaptation among vulnerable communities, implemented jointly by the Department of Agricultural Extension (DAE) and Food and Agriculture Organisation of the UN (FAO) (Baas & Ramasamy, 2008). Project outputs included learning lessons from CBA. The project assessed existing locally specific risk-coping strategies and technologies, monitored local agro-meteorological data and downscaled climate scenarios (Baas & Ramasamy, 2008), intending to create an overlap between local and scientific knowledge (Torres, 2009). Due to the lack of reliable downscaled climate data, pilot projects focused on 'no regrets' options for field testing, such as drought-tolerant crops in the North-West (Baas & Ramasamy, 2008).

In the first pilot phase (2005–2007), mainstreaming and scaling up were not effectively addressed (Baas & Ramasamy, 2008), but in later stages, lessons were learnt and the broad-based reach of DAE's 12,000 agricultural extension workers were tapped (FAO, 2010a). An independent CDMP evaluation found that LACC was successful and proposed further embedding climate-related knowledge in forthcoming projects, but found that gender issues require further attention (Russell, Mahbub, Khan, & Islam, 2009). Political turmoil and staff continuity were also challenges (Luxbacher, 2011). A programme review recognized information and communication gaps, such as illiteracy and the bias towards production of printed materials, the under-utilized role of information and communication technologies (ICT), and the absence of a communications plan (FAO, 2010a). To overcome these pitfalls, a deliberate effort was proposed to mainstream adaptation within national policy and development planning, including advocacy and policy briefs for law-makers and local officials, and better inclusion of communication activities at every level (FAO, 2010a). Collaboration with the Agricultural Information Service (AIS) and inclusion of climate change into education curricula were also recommended (FAO, 2010a). A continuing barrier was the reach of extension workers, with the ratio of extension workers to farmers at 1:12,000 (FAO, 2010a). General lessons were that integration of disaster risk reduction (DRR) and adaptation into operational local-level frameworks are crucial to initiate long-term processes, and there is no need to set up separate institutional structures within sectoral line agencies (FAO, 2010b). Adaptation was highlighted as a social learning process, and inclusive and participatory mechanisms can contribute to this learning (FAO, 2011).

Drawing on lessons from the previous projects, the Disaster and Climate Risk Management in Agriculture (DCRMA) project aims to mainstream disaster and climate risk management in the DAE and strengthen its capacity. There is now collaboration with AIS in disseminating success stories from the grassroots level. In a current project 'Agricultural Adaptation in Climate Risk Prone Areas of Bangladesh', many lessons from the LACC are being built upon, including collaboration with farmer field schools, as well as improving community-based early warning systems and rural communication services (BCCRF, 2013). The project seeks to focus on community-based and field-level adaptive research and participatory extension approaches, as well as community-based seed and grain storage infrastructure, water harvesting and small-scale irrigation, drawing on local knowledge to develop regional agro-ecological databases and community-based DRR plans (DAE, 2013). Replication and scaling up of agricultural adaptation options are intended to occur through farmer clubs and water-

management groups (DAE, 2013). It was argued there is no need to create separate 'climate field schools' because farmer field schools will iteratively adjust to climate-related changes like salinity, but communication may be needed to ensure effective innovation (AEC, 2011).

Overall, the case provides nascent evidence of mainstreaming of climate change into the activities of national agricultural institutions and programmes in Bangladesh, building upon years of progress in collaboration with FAO. Climate is also being integrated into the research priorities of the Bangladesh Agricultural Research Council, particularly the high-priority areas of climatic impacts on fisheries, water resources management, forests and disaster management (Hussain & Iqbal, 2011). However, while CBA approaches are articulated within project documents under the DCRMA, it is too early to explore whether these will be effective in practice. Insufficient capacity and lack of coordination among research scientists, extension workers and farmers remain key challenges for the forthcoming project (Rahman, 2011). Furthermore, biodiversity loss and inadequate use of indigenous knowledge in food-related contexts are additional barriers to adaptation (Mallick, Amin, & Rahman, 2012). Lack of inclusion of farmers in research and general lack of awareness about climate change are also barriers. Mainstreaming adaptation in agriculture is an on-going process in Bangladesh. Proposed ways forward include greater communication efforts, coordination amongst stakeholders and collaboration with existing organizations including farmer field schools.

3.2. *Adapting to climate change in semi-arid environments in Mozambique*

In Mozambique, experiences with mainstreaming CBA show that major challenges include the capacity of farmers' organizations and extension services, farmers' access to markets and coordination across implementing agencies, particularly at local levels. National experiences with mainstreaming adaptation into agriculture were reviewed using secondary reports and literature, with a focus on the three-year (2008–2010) UN Joint Program (UNJP) on Environmental Mainstreaming and Adaptation to Climate Change, which aimed to help Mozambique integrate climate change into national policy and set up pilot adaptation projects. The programme was designed to align with government planning and strategies, create synergies and avoid duplication. Pilot projects were implemented in Chicualacuala, Gaza Province (FAO, 2011). The activities arguably contributed to realization of Mozambique's NAPA (FAO, 2012b), which prioritizes early warning systems, increasing producer capacity and management of water resources. For example, to strengthen early warning systems, the UNJP assisted the government by rehabilitating and re-equipping a weather station and expanding the reach of the Chicualacuala community radio station. In an area where livestock is crucial to livelihoods, a network of trained community animal health workers (CAHW) was established. Taking actions at the community level meant problems could be identified more accurately and locally appropriate preventive measures could be taken.

An independent evaluation found that the establishment of community groups was highly effective, but the project's relevance was reduced by not fully responding to communities' market-related challenges (Eucker & Reichel, 2012). It was recommended that adaptation strategies ought to be flexible, with a greater focus on 'how' results are achieved (process) rather than 'what' is achieved (Eucker & Reichel, 2012). A further study in Gaza showed communities have multiple viable strategies for reducing climate risks, including livestock management and livelihood diversification, which could be expanded and strengthened through a greater service provision by the government, notably weather forecasts and climate information services (Sacramento, Matavel, Basílio, & Bila, 2012). Expert interviews revealed various institutional barriers to mainstreaming adaptation, including lack of human resources, insufficient data, lack of inter-institutional coordination and communication and scarce financial resources (Sietz et al., 2011). In Mozambique, underlying structural issues such as weak markets for agricultural commodities, poor infrastructure and limited access to micro-finance exacerbate difficulties for smallholder farmers (Osbahr, Twyman, Adger, & Thomas, 2008).

Communities involved in the programme identified that the human resources most important for their livelihoods were health, education, farming skills and extension services (farm and veterinary) (FAO, 2012b). Farmers' organizations also provided social capital for adaptation (FAO, 2012b). The National Directorate of Agrarian Extension (DNEA) is the main institution responsible for agricultural extension. It does not have any climate-specific programmes, demonstrating that adaptation is insufficiently mainstreamed at present. However, Mozambique's Third Poverty Reduction Strategy (PARPA III) identified some climate-related activities, including water-management and improved seed varieties. One challenge is that Mozambique's extension services are relatively new, formed only in 1987 amidst a challenging political environment (Gemo, Eicher, & Eclemariam, 2005). It is difficult to finance extension services in subsistence and semi-subsistence economies without taxable agricultural exports (Eicher, 2004). The majority of farmers face challenges accessing extension services, with some 2000 extension workers covering a rural population of over 14 million (FAOSTAT, 2012). This demonstrates a need to scale up existing extension services to increase farmers' food security and resilience. Adaptation is now being integrated in the ASAP-supported value chain development project in

Mozambique, with extension services recognized as a barrier, as well as gaps in financial services and smallholder market access (IFAD, 2012b). Building the capacity of farmers' organizations is another key priority (IFAD, 2012b). Overall, although Mozambique has a supportive national-level legislative environment and awareness among donors is high, there is still limited institutional capacity for mainstreaming initiatives at provincial and district levels (Sietz et al., 2011). In spite of the decentralization process, lack of communication, coordination, funding and poor information dissemination impede mainstreaming adaptation at local levels.

3.3. *Climate-smart adaptation research in Rakai District, Uganda*

In Uganda, the main finding is that non-functional policies and regulations at national or sub-national levels inhibit mainstreaming of CBA. Participatory research has been undertaken by the International Institute of Tropical Agriculture (IITA) with producers in assessment of vulnerability and evaluation of adaptation options, as part of the Climate Change Agriculture and Food Security (CCAFS) research programme. Climate change threatens to decrease yields, reduce farm revenues, worsen food insecurity and deepen rural poverty (Nabikolo, Bashaasha, Mangheni, & Majaliwa, 2012; UNDP, 2013; Waithaka, Nelson, Thomas, & Kyotalimye, 2013). Participatory vulnerability assessments were conducted in Rakai District to capture interactions among biophysical, social, political–institutional, socio-cultural, economic and environmental variables. Twenty focus-groups were conducted in 10 different zones, 2 per zone, separating men and women to capture gendered differences in perceptions. Participatory discussions were held on climatic and environmental changes that have occurred in the last 2–3 decades, changes to farming practices, climate constraints experienced and adaptation practices farmers use to cope with climate challenges. In addition, in-depth key informant interviews were conducted with selected smallholder farmers, political leaders, public extension entities, NGOs and businesses (agro-produce marketers and agro-input dealers). Furthermore, a formal survey that utilized structured questionnaires was administered to individual farmers selected randomly from sites, to complement the analysis and interpretation of findings (Kyazze & Kristjanson, 2011). The research aimed to analyse collected information, and generate and present different climate-smart scenarios to male and female producers to help them evaluate their applicability and sustainability. Smallholder farmers took centre stage in developing 'climate-smart' options on the premise that effective participation of vulnerable communities is likely to enhance design, adoption and ownership of adaptation.

Although what works in a pilot might not necessarily work elsewhere, lessons so far learnt from the project provide useful insights by highlighting success and constraining factors that could be applied by large-scale programmes that intend to scale up CBA. The participation of a broad range of stakeholders enabled shared learning and fostered commitment to undertake actions. Since the climate-smart options agreed upon measure up to realities on the ground, their adoption has relatively higher chances of sustainability. Sustainability strategies have been weaved in right at the start to avoid over-dependence, including identifying the right stakeholders for institutional support, building relevant capacities of different stakeholders, specifying roles for different actors and securing commitment from them to deliver on roles.

Various factors may constrain adoption of climate-smart options if they are not dealt with. In focus group discussions, it emerged that local policies exacerbated farmer vulnerability; for instance in the past farmers had access to communal grazing lands which were utilized during periods of fodder scarcity. Similarly, farmers used to produce crops in wetlands during droughts and return to their upland plots during the rainy season (Turyahabwe, Kakuru, Tweheyo, & Tumusiime, 2013). Yet communal grazing lands and wetlands were leased out by the district land board to a few well-off farmers, who have either fenced them off or used them to establish commercial eucalyptus woodlots (Ampaire, 2013). As a result, poor smallholder farmers no longer have access to these resources. In addition, planting of eucalyptus in the wetlands resulted in lowering of the water table and drying of community wells (Ampaire, 2013). Women and children were particularly affected as they have to travel up to four kilometres to fetch water during the dry season. There was little incentive for adopting 'climate-smart' practices such as agroforestry, as the land is limited and farmers cannot accommodate the time lag on the return on investment.

Based on these insights, a more detailed study was conducted to understand policy formulation and implementation processes and constraints to CBA. A range of national-level policies and regulations currently exist on the paper to guide access to and use of natural resources, but these are almost non-functional at local levels (Rwakakamba, 2009). Examples include the Uganda Forestry Policy (2001), National Environment Act (1998) and National Wetlands Policy (1995). Findings affirm that CBA is constrained by lack of policy implementation, which is brought about by multiple factors, including exclusion of implementers in the formulation process, inadequate knowledge about policies (Glass, 2007), poor coordination among actors, lack of clarity in roles, limited resources and political interference, coupled with corruption. In addition, land tenure is insecure and the Uganda National Land Policy was only recently approved in 2013, following prior allegations of land grabbing

from land owners, media and civil society. At present, smallholders feel helpless as there are no laws implemented to assure their access to land and other natural resources. Climate-smart solutions will only be effective if there is political will among national and local leaders to address constraints jointly. IITA and partners are planning policy engagement actions at national and sub-national levels, and sharing evidence both on technical issues, such as 'no regret' climate-smart technologies in coffee–banana systems, and on institutional issues, such as gender inclusion and resource access. While it is too early to explore whether specific measures will be mainstreamed at national levels, the aim is to enable more inclusive implementation of natural resource use policies.

3.4. Participatory research and mobilization of young farmers in Karnal District, India

The key lesson from India is that CBA can mobilize young farmers and provide a platform for scaling out adaptation technologies. The International Maize and Wheat Improvement Centre (CIMMYT), under the CCAFS research programme, is conducting participatory action research with farmers in Karnal, an agriculturally vibrant region of Haryana state (Aggarwal et al., 2010). This case study draws on primary data including household surveys (Singh, 2013) and participatory technology evaluation trials undertaken in the climate-smart villages, in which data were collected by CIMMYT–CCAFS (Table 1). Since the mid-1960s, increases in agricultural productivity, rapid industrial growth and expansion of the non-formal rural economy have quadrupled per capita Gross Domestic Product (GDP), and markedly reduced poverty. However, securing these gains is becoming a challenge in the context of soaring food and fuel prices, volatile markets, global economic downturn, diversion of human capital from agriculture, soil degradation, shrinking farm sizes, depletion of water resources and overarching effects of climate change (Ambast, Tyagi, & Paul, 2006; Humphreys et al., 2010; Jat et al., 2012). Climate change is projected to lead to uncertain onset of monsoons and more frequent extremes of weather (Aggarwal, Joshi, Ingram, & Gupta, 2004). Also, other competitive sectors and schemes such as 'MNREGA' (Mahatma Gandhi National Rural Employment Guarantee Act) have diverted farm labour. Conventional agricultural technologies, farming practices and linear out-scaling approaches (Swanson, 2008) under emerging climatic risks further exacerbate the challenges and make farming unattractive to farmers in general, and youth and women in particular (GCWA, 2012).

Conservation agriculture (CA) management technologies offer some solutions to the emerging challenges of climate change across the Ganges River basin, including Haryana, by maintaining soil fertility and water-holding capacity in conditions of unpredictable monsoons

Table 1. Climate-smart technologies adopted and disseminated, and their monetary advantages.

	Technologies adapted and disseminated	Climate-smart category	Yield gains over local farmers' practices (kg ha^{-1})	Monetary gains over local farmers' practices (US$ ha^{-1})	Number of farmers who benefited
1	Laser land levelling	Water smart	480	144	250
2	No-till wheat with residue retention (turbo seeder)	Carbon, energy and water smart	600	174	60
3	Direct dry seeded rice	Water and energy smart	00	180	60
4	Site-specific nutrient management, nutrient expert decision support tool rice–wheat system	Nutrient smart	550	127	82
5	GreenSeeker sensor guided nitrogen application	Nutrient smart	275	72	10
6	*Diversification/intensification*				
A	Relay mungbean in wheat	Carbon smart	855	217	5
B	Dual purpose wheat		1230	313	5
C	Introduction of maize replacing rice	Water smart	[a]	315	2

[a]Two different crops, hence comparisons of yield are not made.

(Erenstein, Farooq, Malik, & Sharif, 2008; Gathala et al., 2011; Jat et al., 2009). Out-scaling (replication) of these relatively knowledge-intensive technologies and practices is more difficult than Green Revolution technologies (new seeds, fertilizers and irrigation). Significant efforts are being made on development and dissemination of new technologies through various institutions, but adoption remains slow. Major bottlenecks include the increasing average age of farmers, traditional mindsets, youth moving out of farming, individualistic and linear technology development, adaptation and dissemination.

Discussing with communities the ways to break the impasse, CIMMYT decided to undertake technology development with young farmers in the belief that engaging young farmers in CBA will facilitate adaptation and adoption of new technologies. CIMMYT also recognized the advantage of bringing young farmers together to influence policy-makers to support technology promotion, targeting not only adaptation and mitigation but also improving farm profitability and generating alternate employment for rural youth through technology-led business opportunities. The other perceived benefit was to develop suitable institutional mechanisms for buying and sharing assets such as expensive farm machinery, and for using resources more precisely at community-level. CIMMYT interacted with a group of young farmers from Taraori village, Karnal District. The response was overwhelming. During the interactions, farmer groups showed keen interest in new-generation technologies to resolve problems of seeding rice with less labour, precision in levelling to save irrigation water, residue management towards improving soil fertility and water-holding capacity, eliminating tillage to save on fuel, energy and water, and improving nutrient-use efficiency. All of these actions enhance rural livelihoods by increasing farmers' incomes, thus reducing vulnerability (Aggarwal et al., 2004), and enabling adaptation benefits such as ability to respond rapidly and cost effectively to delayed and unpredictable start of the growing season, or minimizing crop losses during dry periods.

Enthusiasm was so high that a group of 20 young farmers from the village took the initiative to form a society registered as 'Society for Conservation of Natural Resources and Empowering Rural Youth'. Since the inception of this society, policy-makers have visited and interacted with these farmers to learn more about resource-efficient, climate-smart and profitable technologies. Also, as farmers' participation in technology development and adaptation is critical, the CCAFS research programme established a participatory strategic research platform at the village level to serve as a capacity-building awareness-creation platform for different stakeholders. As summarized in Table 1, adaptive technologies were demonstrated and disseminated to a large number of farmers in the local area, and more widely across Haryana. Input and output data were collected from selected farmers using a simple checklist, and subsequently market prices for input costs and crop production were used to calculate net returns.

Farmers of the society have been active in publicizing the technologies through print and electronic media, including local and national newspapers, on television, and participation in various state, national and international level meetings. Recognizing the innovative contributions of these young farmers, the State-level Innovative Farmer Award was presented to them by the Chief Minister of Haryana in December 2012. The Chief Minister also announced state-wide incentives for community-based climate-smart and resource-efficient technologies, primarily CA and resource-efficient mechanization, providing evidence that CBA is being mainstreamed at the sub-national level. The new Haryana State Agriculture Policy, adopted in 2014, recognizes emerging threats due to climate change, and emphasizes adaptive measures to minimize these consequences (ADH, 2014). The fact adaptation actions were community-based captured national media and political attention. This village became a role model for rural youth in five more such young farmer cooperatives. CBA built the capacity of young farmers so they not only adopted new technologies but also provided services to other farmers to earn money. An important element of this was the participatory approach and non-linear flow of information. In traditional extension systems, different organizations work in isolation and often deliver conflicting messages to farmers, but in this approach, farmers formed a common platform to debate and reach consensus on the new technologies.

4. Discussion

There are emerging evidence adaptation programmes and strategies for agriculture and are more likely to be effective if they directly involve communities that are innovating and implementing CBA at local levels. In India, the enthusiasm of young farmers in community-based organizations enabled the adoption, piloting and subsequent wider dissemination of adaptation technologies, overcoming social barriers to adaptation. It is perhaps unsurprising this successful case study has targeted younger farmers, who are most likely to be interested in the long-term future of farming. In Uganda, a participatory approach to assessment of climate-related risks and vulnerabilities and development of profitable climate-smart options captured the priorities and preferences of different categories of local stakeholders and created local ownership. These insights are particularly relevant when we consider the locally specific impacts of climate change and the uncertainty about how impacts may manifest themselves. Local perceptions by farmers about climate can also be matched up with meteorological data at national weather stations, as has been done in Uganda (Osbahr, Dorward, Stern, & Cooper, 2011). Recognizing adaptation as a process of social learning (Collins & Ison, 2009), policy-makers will do well to recognize the value of locally specific knowledge from CBA, through farmer field schools and other means.

The major opportunity to bring social learning on adaptation to the national level is via existing advisory services rather than creation of new networks and institutions. Extension services provide a vital social learning role (Eicher, 2004) and are a medium through which farmers access climate-relevant information, including market information and technologies, and so through which CBA

could potentially be scaled out. But in many cases, including Bangladesh and Mozambique the current reach of extension services is limited. For example, in Mozambique, extension services are available only to a minority of farmers (from 4% in Inhambane to 7% in Maputo province; IFAD, 2012b). Furthermore, extension services usually target relatively wealthier households and thus may not reach the most vulnerable (Cunguara & Moder, 2011), presenting a challenge for scaling up CBA.

Mainstreaming CBA in agriculture faces particular institutional, social, policy, market and financial barriers. In Mozambique, there are barriers to national-level mainstreaming related to non-alignment of policies, strategies and plans, poor institutional coordination, and limited human and financial resources. In Bangladesh, key problems identified were lack of inter-agency coordination, communication barriers and literacy gaps. These barriers to CBA, including communication and literacy gaps, are highlighted in other studies (Spires, Shackleton, & Cundill, 2014). Proposed solutions that may be cost-effective as well as institutionally feasible include: farmers' organizations and low-cost platforms for shared learning on adaptation, particularly across local and national levels; better use of non-print communications media to overcome literacy gaps, in particular verbal communications by radio, mobile phones and face-to-face exchange; as well as tackling market barriers through trade reform, improved transport and storage facilities.

Furthermore, CBA activities implemented in these case studies were not necessarily responding to climate change but also to other challenges, such as conserving fuel and water in India. Climate factors may have less influence than other socio-economic stresses in shaping agricultural livelihoods (Mertz, Mbow, Reenberg, & Diouf, 2009; Ziervogel, Bharwani, & Downing, 2006). Successful adaptation policy in agriculture will thus need to create synergies with agricultural development to enhance adaptive capacity, while recognizing that 'modern farming' can have both positive and negative impacts on adaptive capacity (Dixon, Stringer, & Challinor, 2014).

In the Ugandan case, gaps in the implementation of existing land and forestry legislation made farmers more vulnerable to climate risks. Overall, CBA can be constrained by both lack of policy implementation, and by policy implementation. Existing institutions may not be inclusive or community-focused, limiting the extent to which local-level CBA can mainstreamed (scaled up) into national policies. These findings support the view that scaling up requires an enabling policy, political and institutional space, as well as financial and market space for an initiative to grow (Linn, 2012). Mainstreaming CBA needs to move beyond identifying and promoting best practices, towards tackling drivers of vulnerability and institutionalizing an enabling environment for CBA to occur as a process.

5. Conclusions and implications

This article draws upon large-scale and local-level cases in Bangladesh, Mozambique, Uganda and India to appraise the opportunities and barriers for community-based approaches to provide effective inputs to higher level policy or larger scale programmes, and for large-scale programmes to respond effectively. Extension services and farmers' organizations are highlighted as mechanisms for linking between national-level and community-level adaptation, while success factors include participative and locally driven vulnerability assessments and tailoring of adaptation technologies to local contexts, mapping local institutions and working in partnership across institutions. Barriers include weak governance, gaps in the regulatory and policy environment, high opportunity costs, low literacy and underdeveloped markets.

Mainstreaming CBA in agriculture raises issues of what constitutes 'additionality' in adaptation. Is it sufficient to mainstream treatment of climate risks into existing agricultural development programmes and extension services, or is there need to extend such services in climate-vulnerable areas? The evidence from Bangladesh and Mozambique suggests mainstreaming adaptation into existing services may not be sufficient to reduce vulnerability, and additional investment is required to scale up support. The new ASAP-supported IFAD programme in Mozambique recognizes that access to technology, extension and infrastructure alone does not demonstrably increase household incomes if these efforts are isolated from access to value chains (IFAD, 2012b), and is thus investing in the fundamentals of rural development (infrastructure, market access and information). Implicit in this is an understanding that the key to building adaptive capacity is to address the existing 'development deficit' (Parry et al., 2009). Crucially, mainstreaming must not become a 'ploy by developed countries' to avoid providing additional adaptation finance (Klein, 2010, p. 45).

Since CBA encounters barriers at both national and sub-national levels, approaches or policies may be needed to overcome these gaps at different scales. Further research is needed on policies and reforms to strengthen adaptation as a social learning process, recognizing that it may be necessary to address barriers at broader scales (institutions, regulations or markets) in order to overcome local resource constraints. Creating space for both development partners and farmers to convey their adaptation priorities to policy-makers is a novel and potentially important modality for adaptation mainstreaming in agriculture, again a social learning process (Kristjanson, Harvey, Van Epp, & Thornton, 2014). It may be prudent to consider a broad interpretation of mainstreaming, to include improvement of structural and legal frameworks towards the objective of reducing the underlying vulnerability of all farmers. The experiences reported in this article demonstrate that mainstreaming adaptation in agriculture needs to go beyond 'climate-proofing' agricultural development, towards tackling the underlying drivers of poverty that exacerbate vulnerability and constrain adaptation.

References

Adger, N., Huq, S., Brown, K., Conway, D., & Hulme, M. (2003). Adaptation to climate change in the developing world. *Progress in Development Studies*, *3*(3), 179–195.

ADH. (2014). *Haryana state agriculture policy*. Haryana: Agriculture Department. Retrieved June 16, 2013, from http://agriharyana.nic.in/Agriculture%20Policy/English%20Haryana_State_Agriculture_Policy_Draft.pdf

AEC. (2011). Farmer field schools in the agricultural extension component (AEC). Retrieved September 22, 2013, from http://bangladesh.ipm-info.org/library/documents/aec_ffs_process_documentation.pdf

Aggarwal, P.K., Ayyappan, S., Chand, R., Gupta, H.S., Gupta, R., Jat, M.L., … Virmani, S.M. (2010). State of Indian agriculture – The Indo-Gangetic Plain. *National Academy of Agricultural Sciences*, *56*, Retrieved from http://naasindia.org/documents/Annual%20Report%202010-11.pdf

Aggarwal, P.K., Joshi, P.K., Ingram, J.S.I., & Gupta, R.K. (2004). Adapting food systems of the Indo-Gangetic Plains to global environmental change: Key information needs to improve policy formulation. *Environmental Science & Policy*, *7*(6), 487–498.

Ambast, S.K., Tyagi, N.K., & Paul, S.K. (2006). Management of declining groundwater in the trans-Indo-Gangetic Plain (India): Some options. *Agricultural Water Management*, *82*(3), 279–296.

Ampaire, E. (2013). Unpublished IITA report, internal document.

Baas, S., & Ramasamy, S. (2008). *Community-based adaptation in action*. Environment and Natural Resources Management Series 14. Rome: FAO. Retrieved from http://www.preventionweb.net/english/professional/publications/v.php?id=8311

BCCRF. (2013). Retrieved from http://bccrf-bd.org

Binswanger-Mkhize, H.P., De Regt, J.P., & Spector, S. (2010). *Local and community driven development: Moving to scale in theory and practice. New Frontiers of Social Policy*, Washington, DC: World Bank.

Blaikie, P. (2007). Is small really beautiful? Community-based natural resource management in Malawi and Botswana. *World Development*, *34*(11), 1942–1957.

CARE. (2009). *Mainstreaming climate change adaptation*. Retrieved September 22, 2013, from http://www.careclimatechange.org/files/adaptation/CARE_VN_Mainstreaming_Handbook.pdf

Carter, M.R., Little, P.D., Mogues, T., & Negatu, W. (2007). Poverty traps and natural disasters in Ethiopia and Honduras. *World Development*, *35*(5), 835–856.

Collins, K., & Ison, R. (2009). Living with environmental change: Adaptation as social learning. *Environmental Policy and Governance*, *19*(6), 351–357.

Cunguara, B., & Moder, K. (2011). Is agricultural extension helping the poor? Evidence from rural Mozambique. *Journal of African Economics*, *20*(4), 562–595.

DAE. (2013). *Draft project outline documents*. Unpublished manuscript. Department of Agricultural Extension (DAE).

Dixon, J.L., Stringer, L.C., & Challinor, A.J. (2014). Farming system evolution and adaptive capacity: Insights for adaptation support. *Resources*, *3*(1), 182–214. doi:10.3390/resources3010182

Eicher, K.C. (2004). Mozambique: Building African models of agricultural extension. In W. Rivera & G. Alex (Eds.), *Extension reform for rural development* (pp. 12–19). Agriculture and Rural Development Discussion Paper 12, Washington, DC: World Bank.

Erenstein, O., Farooq, U., Malik, R.K., & Sharif, M. (2008). On-farm impacts of zero tillage wheat in South Asia's rice–wheat systems. *Field Crop Research*, *105*, 240–252.

Eriksen, S., Aldunce, P., Bahinipati, C.S., Martins, R.D., Molefe, J.I., Nhemachena, C., … Ulsrud, K. (2011). When not every response to climate change is a good one: Identifying principles for sustainable adaptation. *Climate and Development*, *3*(1), 7–20. doi:10.3763/cdev.2010.0060

Eucker, D., & Reichel, B. (2012). *Final evaluation: Environment mainstreaming and adaptation to climate change*. MDG Fund. Retrieved from http://www.mdgfund.org/sites/default/files/Mozambique%20-%20Environment%20-%20Final%20Evaluation%20Report.pdf

FAO. (2010a). Communication assessment and action plan for the LACC Project. CSDI Technical Paper, FAO, Rome. Retrieved from http://www.africa-adapt.net/media/resources/467/CSDI.LACC%20Project.pdf

FAO. (2010b). General lessons from livelihood adaption to climate change project. FAO, Rome. Retrieved from http://www.fao.org/climatechange/24647–05e72d0b0d66b38b187f892ca94bdaa3b.pdf

FAO. (2011). Lessons from the field: Experiences from FAO climate change projects. FAO, Rome. Retrieved from http://www.fao.org/docrep/014/i2207e/i2207e.pdf

FAO. (2012a). *Incorporating climate change considerations into agricultural investment programmes: A guidance document*. Rome: FAO Investment Centre. Retrieved from http://www.fao.org/docrep/016/i2778e/i2778e.pdf

FAO. (2012b). Adaptation to climate change in semi-arid environment: Experience and lessons from Mozambique. *Environment and Natural Resource Management*, *19*, Retrieved from http://www.fao.org/docrep/015/i2581e/i2581e00.pdf

FAOSTAT. (2012). Retrieved from http://www.faostat.fao.org

Gathala, M., Ladha, J.K., Balyan, V., Saharawat, Y.S., Kumar, V., & Sharma, P.K. (2011). Effect of tillage and crop establishment methods on physical properties of a medium-textured soil under 7-year rice–wheat rotation. *Soil Science Society of America Journal*, *75*, 1–12.

GCWA. (2012). Proceedings of first Global Conference on Women in Agriculture (GCWA), 13–15 March, 2012, New Delhi, India. Indian Council of Agricultural Research and Asia Pacific Association of Agricultural Research Institutions, p. 82.

Gemo, H., Eicher, C., & Eclemariam, S. (2005). *Mozambique's experience in building a national extension system*. Michigan: Michigan State University Press.

Glass, S. (2007). *Implementing Uganda's national wetlands policy: A case study of Kabale district*. Independent Study Project (ISP) Collection. Paper 101. Retrieved from http://digitalcollections.sit.edu/isp_collection/101

Heltberg, R., Siegel, P., & Jorgensen, S. (2009). Addressing human vulnerability to climate change: Toward a 'no regrets' approach. *Global Environmental Change*, *19*(1), 89–99.

Humphreys, E., Kukal, S.S., Christen, E.W., Hira, G.S., Balwinder-Singh, Sudhir-Yadav, & Sharma, R.K. (2010).

Halting the ground water decline in north-west India – Which crop technologies will be winner. *Advances in Agronomy*, *109*, 155–217.

Huq, S., Rahman, Q., Konate, M., Sokona, Y., & Reid, H. (2003). *Mainstreaming adaptation to climate change in least developed countries (LDCs)*. London: IIED.

Hussain, S., & Iqbal, A. (2011). *Research priorities in Bangladesh agriculture*. Dhaka: Bangladesh Agricultural Research Council.

IFAD. (2012a). Adaptation for smallholder agriculture programme (ASAP). Rome, Italy. Retrieved from http://www.ifad.org/climate/asap/asap.pdf

IFAD. (2012b). Pro-poor value chain development in the Maputo and Limpopo corridors (PROSUL). Retrieved from http://www.ifad.org/operations/projects/design/106/mozambique.pdf

Jat, M.L., Gathala, M.K., Ladha, J.K., Saharawat, Y.S., Jat, A.S., Kumar, V., … Gupta, R. (2009). Evaluation of precision land leveling and double zero-till systems in the rice–wheat rotation: Water use productivity, profitability and soil physical properties. *Soil Tillage Research*, *105*, 112–121.

Jat, M.L., Malik, R.K., Saharawat, Y.S., Gupta, R., Bhag, M., & Paroda, R. (Eds.). (2012). *Regional dialogue on conservation agricultural in South Asia. Proceedings and recommendations, New Delhi, India, 1–2 November 2011*, p. 34.

Kansiime, M. (2012). Community-based adaptation for improved rural livelihoods: A case in eastern Uganda. *Climate and Development*, *4*(4), 275–287. doi:10.1080/17565529.2012.730035

Klein, R.J.T. (2010). Mainstreaming climate adaptation into development: A policy dilemma. In A. Ansohn & B. Pleskovic (Eds.), *Climate governance and development, Berlin Workshop Series 2010* (pp. 35–52). Washington, DC: World Bank.

Kristjanson, P., Harvey, B., Van Epp, M., & Thornton, P.K. (2014). Social learning and sustainable development. *Nature Climate Change*, *4*, 5–7. doi:10.1038/nclimate2080

Kyazze, F.B., & Kristjanson, P. (2011). *Summary of baseline household survey results: Rakai district, south central Uganda*. Copenhagen, Denmark: CGIAR Research Program on Climate Change, Agriculture and Food Security (CCAFS).

Linn, J. (Ed.). (2012). *Scaling up in agricultural, rural development and nutrition*. IFPRI. Retrieved from http://www.ifpri.org/publication/scaling-agriculture-rural-development-and-nutrition

Luxbacher, K. (2011). *Bangladesh' comprehensive disaster management programme*. Inside Stories: CDKN.

Mallick, D., Amin, A., & Rahman, A. (2012). *Case study on climate compatible development in agriculture for food security in Bangladesh*. Bangladesh Centre for Advanced Studies (BCAS), Dhaka, September 2012. Retrieved from https://germanwatch.org/en/download/8347.pdf

Mansuri, G., & Rao, V. (2004). Community-based and -driven development: A critical review. *The World Bank Research Observer*, *19*(1), 1–39.

Mertz, O., Mbow, C., Reenberg, A., & Diouf, A. (2009). Farmers perceptions of climate change and agricultural adaptation strategies in rural Sahel. *Environmental Management*, *43*, 804–816.

Nabikolo, D., Bashaasha, B., Mangheni, M., & Majaliwa, J.G.M. (2012). Determinants of climate change adaptation among male and female headed farm households in eastern Uganda. *African Crop Science Journal*, *20*(2), 203–212.

Osbahr, H., Dorward, P., Stern, R., & Cooper, S. (2011). Supporting agricultural innovation in Uganda to respond to climate risk: Linking climate change and variability with farmer perceptions. *Experimental Agriculture*, *47*, 293–316. doi:10.1017/s0014479710000785

Osbahr, H., Twyman, C., Adger, W.N., & Thomas, D.S.G. (2008). Effective livelihood adaptation to climate change disturbance: Scale dimensions of practice in Mozambique. *Geoforum*, *39*(6), 1951–1964.

Parry, M., Arnell, N., Berry, P., Dodman, D., Fankhauser, S., Hope, C., ... Wheeler, T. (2009). *Assessing the costs of adaptation to climate change*. London: IIED.

Rahman, M.M. (2011). Country report: Bangladesh, ADBI-APO workshop on climate change and its impact on agriculture, Seoul, Republic of Korea, 13–16 December 2011.

Reed, M.S., Graves, A., Dandy, N., Posthumus, H., Hubacek, K., Morris, J., ... Stringer, L.C. (2009). Who's in and why? A typology of stakeholder analysis methods for natural resource management. *Journal of Environmental Management*, *90*, 1933–1949.

Reid, H., Alam, M., Berger, R., Cannon, T., Huq, S., & Milligan, A. (2009). Community-based adaptation to climate change: An overview. In Community-based adaptation to climate change, Participatory Learning and Action (PLA) 60 (pp. 11–33). London: IIED. Retrieved from http://pubs.iied.org/pdfs/14573IIED.pdf

Russell, N., Mahbub, A. Q. M., Khan, M.H., & Islam, N. (2009). *Bangladesh comprehensive disaster management programme: Terminal evaluation*. Retrieved from https://www.climate-eval.org/sites/default/files/evaluations/371%20Comprehensive%20Disaster%20Management%20Program.pdf

Rwakakamba, T. (2009). How effective are Uganda's environmental policies? *Mountain Research and Development*, *29*(2), 121–127.

Sacramento, A., Matavel, A., Basílio, M., & Bila, S. (2012). Climate change impacts and coping strategies in Chicualacuala district. Gaza Province: UNEP. Retrieved from http://www.unep.org/climatechange/adaptation/Portals/133/documents/Chicualacuala_Report-Climate_Change_Impacts_n_Coping_Strategies.pdf

Sietz, D., Boschutzz, M., & Klein, R. J. T. (2011). Mainstreaming climate adaptation into development assistance: Rationale, institutional barriers and opportunities in Mozambique. *Environmental Science & Policy*, *14*(4), 493–502.

Singh, R.K.P. (2013). *Summary of baseline household survey results*. May 2013, CCAFS. Retrieved from http://ccafs.cgiar.org/publications/summary-baseline-household-survey-results-karnal-harayana-state-india

Spires, M., Shackleton, S., & Cundill, G. (2014). Barriers to implementing planned community-based adaptation in developing countries: A systematic literature review. *Climate and Development*, *6*(3), 277–287. doi:10.1080/17565529.2014.886995

Stringer, L.C., Mkwambisi, D.D., Dougill, A.J., & Dyer, J.C. (2010). Adaptation to climate change and desertification: Perspectives from national policy and autonomous practice in Malawi. *Climate and Development*, *2*, 145–160.

Swanson, B. (2008). *Global review of good agricultural extension and advisory service practices*. Rome: FAO.

Taylor, M. (2003). *Public policy in the community*. Hampshire: Palgrave Macmillan.

Thomas, D., & Twyman, C. (2005). Equity and justice in climate change adaptation amongst natural-resource-dependent societies. *Global Environmental Change*, *15*(2), 115–124.

Torres, C. (2009). *Advancing adaptation through communication for development*. Rome: FAO. Retrieved from http://www.fao.org/docrep/012/i1553e/i1553e00.pdf

Turyahabwe, N., Kakuru, W., Tweheyo, M., & Tumusiime, D.M. (2013). Contribution of wetland resources to household food security in Uganda. *Agriculture & Food Security*, *2*(1), 5. doi:10.1186/2048-7010-2-5.

UNDP. (2013). *Climate risk management for sustainable crop production in Uganda: Rakai and Kapchorwa districts*. New York, NY: United Nations Development Programme (UNDP), Bureau for Crisis Prevention and Recovery (BCPR).

UNDP-UNEP. (2011). *Mainstreaming climate change adaptation into development planning: A guide for practitioners*. Retrieved from http://www.unep.org/pdf/mainstreaming-cc-adaptation-web.pdf

UNFCCC. (2011). Decision CP.17. FCCC/CP/2011/9/Add.1. United Nations Framework Convention on Climate Change. Retrieved from http://unfccc.int/files/adaptation/cancun_adaptation_framework/national_adaptation_plans/application/pdf/decision_5_cp_17.pdf

Vermeulen, S.J., Campbell, B.M., & Ingram, J.S.I. (2012). Climate change and food systems. *Annual Review of Environment and Resources*, *37*, 195–222.

Vincent, K., Cull, T., Chanika, D., Hamazakaza, P., Joubert, A., Macome, E., & Mutonhodza-Davies, C. (2013). Farmers' responses to climate variability and change in southern Africa – Is it coping or adaptation? *Climate and Development*, *5*(3), 194–205.

Waithaka, M., Nelson, G.C., Thomas, T.S., & Kyotalimye, M. (2013). *East African agriculture and climate change*. Washington, DC: IFPRI.

World Bank. (2010). *Mainstreaming adaptation to climate change in agriculture and natural resources management projects*. Washington, DC: World Bank. Retrieved from http://preventionweb.net/go/17066

Ziervogel, G., Bharwani, S., & Downing, T.E. (2006). Adapting to climate variability: Pumpkins, people and policy. *Natural Resources Forum*, *30*, 294–305.

CASE STUDY

Gender-sensitive adaptation policy-making in Bangladesh: status and ways forward for improved mainstreaming

Dalia Shabib and Shusmita Khan

Bangladesh

Bangladesh is particularly vulnerable to the impacts of climate change such as flooding, cyclones and drought. Women in Bangladesh are disproportionately affected by these impacts due to the nature of their livelihoods, their social obligations and confines, and their unique nutritional and health requirements, particularly during pregnancy and breastfeeding. Climate change policy in Bangladesh seeks to replicate adaptation policies under the United Nations Framework Convention on Climate Change. This paper will briefly review the policy response to climate change in Bangladesh. As climate adaptation requires a multi-sectoral response, relevant policy concerned with climate, adaptation, poverty, gender and health will be studied. This assessment will determine whether gender issues related to adaptation are addressed in key policy pieces in Bangladesh. Key interventions related to climate change will also be assessed to determine whether gender is integrated in operational activity. Finally, the role of women in the development of adaptation policy will be assessed by outlining their participation in adaptation discourse. Findings indicate that gender-sensitive policies are quite limited. Policies may acknowledge the particular vulnerabilities of women, but operational planning to address these is absent. Whilst some operational responses superficially acknowledge vulnerability and may include women in planning processes, few address the unique impacts of climate change on women.

1. Background

Flooding is a principal climatic hazard in Bangladesh. This annual occurrence is symbiotic with rural livelihoods and food production. The seasonal flooding cycle brings fertility to fields and interconnections between water bodies and waterways to nurture the proliferation of fish. Flooding can also be devastating to households which lose assets and sometimes land due to erosion (Cannon, 2009). Flooding disproportionately affects the poor, forcing them into a cycle of poverty due to their inability to cope with losses and access resources used for housing and livelihoods.

Although the annual flooding cycle has positive effects on rural livelihoods and agriculture, expected trends may disturb this balance. The global climate has changed by nearly 1°C in the past century; Bangladesh has experienced a warming of its climate between 0.4°C and 1.8°C. This will likely continue and perhaps accelerate. Based on the Hadley Centre climate models, predict a 1–2°C warming over the next 50 years. This may exceed what can be expected with natural variability within the next two decades (Adger, Huq, Brown, Conway, & Hulme, 2003). As a result, flooding and other climatic hazards affecting

Bangladesh will likely increase in frequency, intensity and duration. Both monsoon rains and Himalayan headwater flows will increase, swelling the river systems and amplifying flooding risks. Subsequent health risks include increased vector and waterborne diseases. Warming will also intensify dry season effects, namely drought. This will have adverse effects on poor households who cannot afford irrigation. This will increase food insecurity and poor nutrition (International Monetary Fund [IMF], 2013). Rising sea levels mean that the coastline will retreat about 10 km (i.e. 18% of the nation's land area) within the next century. Climate change impacts will also increase arsenic contamination from flooding, waterlogging and salt water intrusion, all of which have detrimental effects on health and agriculture (Cannon, 2009; Rahman et al., 2007; WHO, 2011b).

This paper will outline the specific vulnerabilities that women face in Bangladesh due to climate change, and seek to assess if gender-based considerations are integrated into policy and programming responses. This will be done through an assessment of climate change policy, programming and discourse. The major impetus for Climate Change policy in Bangladesh is the commitment to the United

Nations Framework Convention on Climate Change (UNFCCC). Thus, a brief overview of the key policy pieces steaming from the UNFCCC and the subsequent National Adaptation Programmes of Action (NAPAs) will be provided. This is followed with an assessment of Bangladesh's key operational activity. Finally, an assessment on climate change and adaptation discourse will be provided in order to establish whether discourse includes the voices of women.

2. Gender-based vulnerability to climate change

Climatic hazards have gender-specific effects. This is due to social institutions, behavioural norms and the physiological attributes that leave women more vulnerable to the effects of climate change then men.

2.1. *Social vulnerability*

According to the literature on the 'feminization of poverty', women are more vulnerable to poverty. Studies on income disparity between male- and female-headed households, holding all other things equal (i.e. age, race, sex and education), revealed significant correlations between gender and poverty (Pearce, 1978). Broader approaches to 'human poverty', which include 'access to opportunities' and 'choice' as determinants of poverty, also revealed disparities between genders (Fukuda-Parr, 1999). This lexicon has driven international and national dialogue on poverty alleviation, for example, the Fourth World Conference on Women in Beijing and National Poverty Alleviation Policy in Chile (Cagatay, 1998). Poverty leaves individuals more vulnerable to the impact of climate change, thereby the 12.8% of households in Bangladesh that are female-headed (National Institute of Population Research and Training et al., 2011) are particularly vulnerable to climate change.

Social constructs can also increase the vulnerability of women by leaving them less equipped to respond to the effects of disaster, due to their limited mobility. During flooding, many women fear social retribution for leaving their homes or taking shelter with other men, so seek refuge too late. Their mobility is also hindered by children, dress and fewer women know how to swim. During one cyclone in 1991, 90% of fatalities were women and children. Death rates among people aged 20–44 were 71 per 1000 people for women, compared to 15 per 1000 people for men (WHO, 2011a).

The aftermath of natural disasters may also disproportionately affect women due to the nature of their livelihoods and role in production. Sources of income for many women in the developing world include livestock rearing, collecting water, biomass for fuels, harvesting seedlings and in some cases collecting non-timber forest products. Fuel shortages following climatic events mean biomass is particularly important for cooking food and heating, but burning biomass has adverse health effects as it increases the risks of pulmonary disease and damages the respiratory function. Livestock and biodiversity losses can affect the livelihoods, nutrition and health of women. Furthermore, fetching water becomes more difficult and exposes women to waterborne illnesses. Degraded water systems increase the prevalence of cholera, diarrhoea, dysentery, malaria and typhoid. Higher soil salinity can lead to hypertension and involuntary foetus abortion. In times of drought, women will have to walk longer distances to access water and biomass for fuel. These extra burdens obstruct their pursuit of education and health care (Rahman et al., 2007; WHO, 2011b).

The World Disasters Report (2005) recognizes that women and girls are at higher risk of sexual violence, exploitation and abuse, trafficking and domestic violence during disasters (International Federation of Red Cross and Red Crescent Societies, 2005). Women subjected to violence prior to disasters are more likely to experience increased violence after them and may become separated from their family, friends and other potential support and protective systems. After a disaster, women are more likely to become the victims of domestic and sexual violence and may avoid using shelters as a result of fear (WHO, 2011a).

2.2. *Physiological vulnerability and access to health care*

Individual nutritional status partly determines the ability to cope with the effects of disasters. Women are more prone to nutritional deficiencies because of their unique nutritional needs, especially when pregnant or breastfeeding. In Bangladesh, 36% of non-pregnant adult mothers have chronic energy deficiency (a body mass index of less than 18.5 kg/m) (Bangladesh Bureau of Statistics and UNICEF, 2007). Pregnant and lactating women face additional challenges, as they have greater food and water needs. At any given time, an average of 18–20% of the population of reproductive age is either pregnant or lactating. This increases vulnerability within a group that is already at risk (WHO, 2011a).

The risk of iron deficiency/anaemia is high among women in Bangladesh. In 2004, the prevalence of anaemia among pregnant women was 46%. Anaemia has detrimental effects on pregnancy and delivery as it increases rates of intrauterine growth retardation, low birth weight and perinatal mortality (Micronutrient Initiative). According to the Food and Agriculture Organization, where iron deficiency is prevalent, the risk of women dying at childbirth can increase by as much as 20% (WHO, 2011a).

Inequities in intra-household food distribution may also be exacerbated in the event of a disaster. In Bangladesh, intra-household utilization of food discriminates against women and children. Men often consume food in surplus of their calorific requirements, while women and children receive less than what they require (World Food Program, 2012).

Finally, health services suffer in times of disaster. Climate change thus leads to increases in maternal and infant mortality, and reduced immunization services, family planning services and reproductive health care. Risks associated with delivery increase in the event of disaster. For example, saline water increases hypertension and thus the risk of premature abortion and post-partum haemorrhage (WHO, 2011a). Furthermore, disaster zones and relief shelters usually lack hygienic supplies for pregnant, lactating and menstruating women (Arora-Jonsson, 2011).

3. Climate change and adaptation policies

In assessing adaptation to climate change, it is important to understand individual vulnerability: exposure to external risk, shock, stresses and the ability to cope (Davidson, 1993). As women are more vulnerable to climatic risks, policies and strategies that seek to alleviate vulnerability to climate change must consider gender-specific issues (Nelson et al., 2002).

In order to adequately assess Climate Change policy in Bangladesh, the following portion will provide a brief overview of key policies stemming from the UNFCCC and NAPA, which are the driving force behind domestic climate change policy.

3.1. Methodology of review

The objective of this review is to provide a brief synopsis of key Climate Change and Adaptation policy in Bangladesh. This review will seek to locate whether gender-based approaches are adequately considered. To achieve the objective of the study, the following general methodology was taken:

(1) collecting existing literature on climate change policy for Bangladesh, available both nationally and internationally, to the extent possible,
(2) briefly highlighting key policy that drive operational activity in Bangladesh, as reported in literature and
(3) highlighting policy limitations concerned with gender.

3.2. Overview: climate change adaptation policy in Bangladesh

In June 1992, Bangladesh became one of the 165 signatories of the UNFCCC. In April 1994, the convention was ratified and entered into force the following June.

Subsequent negotiations discussed how to achieve the treaty's aims and emissions targets were established. The Kyoto Protocol (1997) set mitigation targets of lowering of emission to an average of 5% below the 1990 level. The adaptation process then began with Bali Action Plan in 2007, which sought to set targets according to Nationally Appropriate Mitigation Actions. Signatories, including Bangladesh, submitted mitigation targets to the UNFCCC secretariat. Subsequent to the Bali Action Plan, Bangladesh prepared the first Bali Submissions, the NAPA in 2005, which evolved into the Bangladesh Climate Change Strategy and Action Plan (BCCSAP) in 2009 (UNFCCC, 2013).

As a Least Developed Country (LDC), Bangladesh is exempt from responsibility to reduce GHG emissions. Furthermore, under the UNFCCC, a special LDC Fund (LDCF) was created to support climate change adaptation. The LDCF has financed the preparation and implementation of national adaptation programmes in developing countries using revenue from the sale of Certified Emissions Reductions from Clean Development Mechanism projects, in addition to contributions from national donor agencies. National Adaptation Programmes of Action (NAPAs) identify priority action to respond to urgent and immediate needs to adapt to climate change (Konate & Sokona, 2003). In 2005, the NAPA was established in Bangladesh led by the Ministry of Environment and Forests (MOEF), in close consultation with the prime minister and other officials.

In 2009, the prime minister of Bangladesh announced the government's full commitment to the NAPA, by implementing the directives of the BCCSAP through various ministries, departments, local governments, civil society and other stakeholders. The BCCSAP takes the Bangladesh submission on Bali Road Map, particularly the '4 securities', as the starting point and develops a sustainable development strategy centred on climate change. These '4 securities' are food security, water security, energy security and livelihood security (including health) (IMF, 2013). Within the BCCSAP, there are six themes for intervention: Food Security Social Protection and Health, Comprehensive Disaster Management, Infrastructure, Research and Knowledge Management, Mitigation and Low Carbon Development, and Capacity Building and Institutional Strengthening, but costing priorities in Bangladesh are not set beyond initial infrastructure spending ($5 billion over 5 years).

National Policy for climate change and adaptation in Bangladesh is integrated into broad strategies for development and poverty alleviation, such as the Sixth Five-Year Plan and the Annual Development Programme. This was done based on a broad consensus that the environment and production systems should be tailored to minimize damage to the environment to ensure the sustainability of development (IMF, 2013).

4. Is climate change and adaptation policy in Bangladesh gender-sensitive?

As the NAPA was the major impetus to mobilize Climate Change and Adaptation Policy in Bangladesh, it is important to consider the structural mechanisms which ensure gender is integrated in nationally implemented adaptation plans. Through the NAPAs, the UNFCCC recognizes that men and women have different roles in securing livelihoods in the developing world. The Adaptation Fund supports activities that are demonstrably gender-sensitive, and funding proposals are reviewed to ensure that they support participatory processes and monitoring in projects/programmes that disaggregates data on specific indicators by gender (UNFPA and WEDO, 2009; UNDP, 2010).

In Bangladesh, Climate Change Adaptation is incorporated into broad multi-sectoral policies such as development and poverty elevation policies. Thus, the Sixth Five-Year Plan, the Annual Development Programme and the Poverty Reduction Strategy paper were reviewed to assess gender sensitivity. Furthermore, as the directives of these overarching pieces are to be incorporated into sector-specific policy regarding environment, health and gender, the following policies were reviewed: the BCCSAP (2010), the National Environment Policy (1992), the National Plan for Disaster Management 2010–2015 (2010), the Health Policy (2011), the National Women Development Policy (2011) and the Bangladesh National Strategy for Maternal Health (2001).

Findings reveal that the climate change and differentiated gender impacts are not recognized. Policy objectives related to climate change and gender are generally addressed as mutually exclusive issues; existing policies do not consider gender-specific operational activities. The only policy that includes gender-based considerations for climate adaptation is the Poverty Reduction Strategy paper (2005), however operational activities under this policy have not yet fulfilled these objectives. Climate change policies generally recognize women as vulnerable but operational responses are not established. In policies related to gender and reproductive health, the impact of climate change is not considered, with the exception of the National Women Development Policy, which briefly promotes women's role in environmental management and the importance of ensuring facilities for and the security of pregnant women in the event of natural disasters.

5. Climate change and adaptation programming

In order to establish whether Climate Change and Adaptation Programming are gender-sensitive, the following portion will briefly outline the mechanisms for mobilising operational activity through two key programmes: the Comprehensive Disaster Management Program (CDMP) under the Ministry of Food and Disaster Management

(MoFDM), and the Reducing Vulnerability to Climate Change (RVCC) partnership.

5.1. Methodology of review

The CDMP and RVCC initiatives were selected for review as they both seek to coordinate the activities of various actors. The mandate of the CDMP is to ensure relevant government agencies (approximately 35 partner agencies) meet key joint objectives and the RVCC coordinates the activities of NGOs or local bodies at the grass roots level (Independent Commission for Aid Impact, 2011; Klein et al., 2007).

5.2. Overview: climate change adaptation programming in Bangladesh

The NAPA for Bangladesh was established in 2005. It identified 15 priority activities, these include general awareness raising, technical capacity building and implementation of projects in vulnerable areas.

The assumption was that key agencies from different sectors would undertake their own initiatives to mainstream adaptation to climate change into their sectoral programme and plans. However, this has not been successful (Rahman et al., 2007). Furthermore, the MOEF is weak and under-resourced, and thus has little power to implement change. Two major channels for operational activity which reaches the targeted communities include CDMP under the MoFDM and the RVCC partnership.

RVCC is a partnership led by CARE in association with 16 local organizations and two research partners. It aims to reduce vulnerability at the union level and conducts advocacy at multiple levels for appropriate government action. Community-based solutions include construction of houses on raised plinths, floating vegetable gardens, cages for fish and awareness building (CARE, 2006).

CDMP (2003–2007) is a large-scale project that attempts to establish preparedness and response measures at policy and community levels. This is done through capacity building of the MoFDM, partnership development and mainstreaming, enhancing community empowerment, risk reduction across a broad range of hazards and response management. In CDMP II (2010–2014), objectives are increasingly focused on community-based mechanisms such as awareness and response strategies among urban and rural populations, improving coordination capacity of response schemes and community-based adaptation strategies, as well as increased funding and coordination at the ministerial level (Government of Bangladesh, 2008; Ministry of Disaster Management and Relief, 2008).

Bangladesh has many additional activities concerned with climate change conducted by NGOs. These are often siloed efforts involving interventions relating to land-use change, food security, water and biomass energy

(Rahman et al., 2007; Thomalla, Cannon, Huq, Klein, & Schaerer, 2005).

5.3. *Is climate change and adaptation programming, in Bangladesh gender-sensitive?*

Under the RVCC, gender issues were promoted through the participation of women in the management committees. Furthermore in monitoring schemes there are mechanisms to ensure equitable outcomes, particularly in livelihood activities. Thus, the unique role of women in livelihood activities is recognized under the RVCC, but their unique health needs due to the impact of climate change is not addressed.

In independent NGO activity and top-down initiatives such as the CDMP and CDMP II, gender is excluded from operational plans. This indicates the need to integrate gender into the culture and lexicon of Climate Change and Adaptation Programming.

6. Community involvement in climate change adaptation dialogue

Participation is acknowledged as central for ensuring adaptation activities are legitimate (Adger et al., 2005; Reid et al., 2009); furthermore, the UNESCAP policy evaluation framework (UNESCAP, 2003) calls for community acceptance, so the extent to which adaptation planning dialogues include women is also assessed.

Debates on the democratic nature of policy-making must incorporate the interests of all legitimate stakeholders. Adaptation policy must be integrated to activity at the national-, Upazila- and local-level governments, besides the private sector, NGOs and community members. The three sectors that are most active in the implementation of climate change activities are water, agriculture and infrastructure and industry. If gender is not integrated into the objectives and activities of these groups, the inequity of adverse effects will increase. Multi-sectoral responses must thus negotiate the interests of individuals, groups and sectors, and aim to reduce the impact of disasters experienced by different stakeholder groups, with consideration to gender as neglecting this component of operational activity would further exacerbate vulnerability faced by women.

Typically, environmental policy is drafted by a select few decision makers and technocrats. Once initial drafts are complete, consultation may then be extended to include additional stakeholders. For example, the first two versions (1992 and 1993) of the National Environment Management Action Plan (NEMAP) were developed by national consultants with the assistance of the UNDP. Public consultation was carried out in the third round (1994). 'Peoples opinions' where considered; it was decided that 23 grassroots workshops in which government agencies and NGO, including the Coalition of Environmental NGOs (CEN) and the Association of Development Agencies of Bangladesh (ADAB), the Bangladesh Centre for Advanced Studies (BCAS) and the Forum of Environmental Journalists of Bangladesh (FEJB), among others. The format of the workshops was carefully designed to ensure balance of women, farmers, fishermen, officials, academics, NGO workers, businesses, elected representatives and others. Major issues identified in the public consultation were sanitation, health, deforestation, pollution, natural disaster, water and flood control/drainage and agro-chemicals. Gender-specific issues related to climate change were not tabled as priority (Ministry of Environment and Forest, 1995).

The most notable initiative that fosters multi-sectoral responses is the NAPA. This was prepared by the MOEF. Stakeholders involved in preparing the NAPA included government policy-makers, local government representatives, members of the scientific community, researchers, academics, teachers, lawyers, doctors, ethnic groups, media, NGO and CBO representatives, and indigenous women. Members of these stakeholder groups attended the inception workshop, regional stakeholder consultation workshop and the national stakeholder consultation workshop (Rahman et al., 2007). Six sectoral working groups were established, one of which was the 'livelihood, gender and local governance and food security' group coordinated by the Bangladesh Institute of Development Studies. Although it is important for women to be present in all discussions and for gender-based issues to be considered under the six established working group activities, it is unclear if women's groups or gender-based issues were considered in deliberations.

In recent years there has been a shift in approach from top-down initiatives in the 1960–1980s such as the Bangladesh Flood Action Plan, to current community-based approaches exemplified by the establishment of the Disaster Management Bureau under the Ministry of Disaster Management and Relief, which seeks to coordinate the efforts of civil society and NGOs. A number of NGO networks have been established such as the Network for Information, Response and Preparedness Activities in Disaster (2011), and the National Alliance for Risk Reduction and Response Initiative comprising eight international NGOs, which is currently implementing several projects funded by funded by the European Commission and UK Aid (Thomalla et al., 2005). However, information deficits remain and collaboration on gender-specific issues is weak. Furthermore, the membership of these two networks does not include women's groups, which indicates that gender-based considerations may not be prioritized.

7. Conclusions

The unique vulnerabilities and needs of women are poorly mainstreamed into climate change policies and operational activity in Bangladesh. Although women are included in adaptation discourse, gender-specific considerations are seemingly not promoted. Multi-sectoral responses and collaborations to address climate change, thus policies relating to disaster preparedness and response, livelihoods and health care may consider integrating gender issues into priorities. Women must also be involved in both the design and implementation of emerging adaptation activities. Adaptation activity must include ensuring protection from gender-based violence, sufficient food consumption, and access to family planning, antenatal care, safe delivery services and postnatal care. Such mainstreaming of gender issues would help ensure that policy and operational responses to climate change go further in terms of meeting the needs of one of the most vulnerable sectors of society in Bangladesh.

References

Adger, W.N., Arnell, N.W., & Tompkins, E.L. (2005). Successful adaptation to climate change across scales. *Global Environmental Change, 15*(2), 77–86.

Adger, W.N., Huq, S., Brown, K., Conway, D., & Hulme, M. (2003). Adaptation to climate change in the developing world. *Progress in Development Studies, 3*(3), 179–195.

Arora-Jonsson, S. (2011). Virtue and vulnerability: Discourses on women, gender and climate change. *Global Environmental Change. 21*(2), 744–751.

Bangladesh Bureau of Statistics and UNICEF. (2007). *Child and mother nutrition survey of Bangladesh 2005*. Dhaka, Bangladesh: Author.

Cagatay, N. (1998). *Social development and poverty elimination: Gender and poverty*. Social Development and Poverty Elimination Division, Bureau for Development Policy, United Nations Development Programme. New York, USA.

Cannon, T. (2009). Gender and climate hazards in Bangladesh. In G. Terry (Ed.), *Practical action publishing ltd and Oxfam*. Bourton on Dunsmore, Rugby, Warwick, UK: Schumacher Centre for Technology and Development.

CARE. *The reducing vulnerability to climate change (RVCC) project: Final report-September 2006*. Retrieved September 10, 2013, from http://www.carebangladesh.org/publication/Publication_4406527.pdf

Davidson. (1993). In women's relationship with the environment in gender and development. In N. Booram (Ed.), *Centre for gender and development studies* (1(1), pp. 5–10). St. Augustine: The University of the West Indies.

Fukuda-Parr, S. (1999). What does feminization of poverty mean? It isn't just a lack of income. *Feminist Economics, 5*(2), 99–103.

Government of Bangladesh. (2008). *Comprehensive Disaster Management Programme: Phase II Design*. UNICEF, DFID and EU, Dhaka, Bangladesh.

Independent Commission for Aid Impact. (2011). *The Department for International Development's climate change programme in Bangladesh*.

International Federation of Red Cross and Red Crescent Societies. (2005). *World disasters report 2005*. Geneva, Switzerland. ISBN 92-9139-109-3.

International Monetary Fund. (2013). *Bangladesh: Poverty reduction strategy paper*. IMF Country Report No. 13/63. Washington, USA.

Klein, R.J.T., Eriksen, S.E.H., Naess, L.O., Hammill, A., Tanner, T.M., Robledo, C., & O'Brein, K.L. (2007). Portfolio screening to support the mainstreaming of adaptation to climate change into development assistance. *Tyndall Centre for Climate Change Research, 84*(1), 23–44. doi:10.1007/s10584-007-9268-x

Konate, A.M., & Sokona, Y. (2003). *Mainstreaming adaptation to climate change in Least Developed Countries* (LDCs). London: International Institute for Environment and Development (IIED).

Micronutrient Initiative. State of Health and Development in Bangladesh. Retrieved June 4, 2013 from http://www.micronutrient.org/english/view.asp?x=602

Ministry of Disaster Management and Relief. (2008). Comprehensive Disaster Management Programme (CDMP II). Retrieved September 12, 2013, from http://www.cdmp.org.bd/index.php

Ministry of Environment and Forest. (1995). *National Environment Management Action Plan (NEMAP) Volume II: Main Report*. Government of the People's Republic of Bangladesh. Dhaka, Bangladesh

Ministry of Environment and Forest. (2005). *National Adaptation Program of Action (NAPA)*.Government of the People's Republic of Bangladesh. Dhaka, Bangladesh

National Alliance for Risk Reduction and Response Initiatives. Retrieved July 31, 2013, from http://narri-bd.org

National Institute of Population Research and Training, Mitra and Associates, Measure DHS, ICF International. (2011). *Bangladesh Demographic and Health Survey*. USAID, NIPORT, MA. Dhaka, Government of Bangladesh.

Nelson, V., Meadows, K., Cannon, T., Morton, J., & Martin, A. (2002). Uncertain prediction, invisible impacts, and the need to mainstream gender in climate change adaptations. In R. Masika (Ed.), *Gender and development* (10(2) pp. 51–59). Oxford: Routledge.

Network for Information Response and Preparedness Activities in Disaster. (2011). *Collaboration*. Retrieved July 31, 2013, from http://nirapad.org

Pearce, D. (1978). The feminization of poverty: Women, work and welfare. *The Urban & Social Review: Special Issue on Women and Work, 11*(1& 2), 28–37.

Rahman, A.A., Alam, M., Alam, S.S., Uzzaman, R., Rashid, M., & Rabbani, G. (2007). *UNDP Human Development Report 2007/08: Background paper on risks, Vulnerability and Adaptation in Bangladesh*. Bangladesh Centre for Advanced Studies and UNDP.

Reid, H., Alam, M., Berger, R., Cannon, T., Huq, S., & Milligan, A. (2009). Community-based adaptation to climate change: An overview. In H. Ashley, N. Kenton, and A. Milligan (Eds.), *Participatory Learning and Action*, (Vol. 60, pp. 11–33). London: IIED. Community-based adaptation to climate change.

Thomalla, F., Cannon, T., Huq, S., Klein, A.J.T., & Schaerer, C. (2005). Mainstreaming adaptation to climate change in coastal Bangladesh by building civil society alliances. *International Institute for Environment and Development, 176*(67), 668–684.

UNESCAP criteria for evaluating policies or measures- integrating environmental considerations into economic policy making processes. Retrieved August 29, 2013, from http://www.unescap.org/drpad/vc/orientation/M5_6.htm

UNFCCC-Bangladesh experiences with the NAPA process. (2013). Retrieved December 1, 2013, from http://unfccc.int/adaptation/knowledge_resources/ldc_portal/bpll/items/6497.php

United Nations Development Project. (2010). Adaptation fund: Exploring the gender dimensions of climate finance mechanisms. *Global Gender and Climate Alliance.*

United Nations Population Fund (UNFPA) and Women's Environment and Development Organization (WEDO). (2009). *Making NAPAs Work for Women.* UNFPA and WEDO.

World Food Program. (2012). *Nutrition strategy.* Dhaka, Bangladesh.

World Health Organization. (2011a). *Gender, Climate changes and health.*

World Health Organization. (2011b). *The social dimensions of climate change.*

RESEARCH ARTICLE

Power and differential climate change vulnerability among extremely poor people in Northwest Bangladesh: lessons for mainstreaming

Cristina Coirolo[a] and Atiq Rahman[b]

[a]Institute of Development Studies (IDS), University of Sussex, Brighton, UK; [b]Bangladesh Centre for Advanced Studies (BCAS), Dhaka, Bangladesh

This article draws on research findings from fieldwork undertaken in Gaibandha District of Northwest Bangladesh from 2009 to 2010 to analyse the influence of power-related factors on climate change vulnerability and adaptation in two rural communities. The principal aim of the research was to explore the factors that shape *differentiated* vulnerability and adaptive capacity within the two communities, with a focus on extremely poor community members. Findings indicate that climate-related vulnerability is differentiated at the sub-community level, both among different socio-economic, livelihood and social groups, as well as within them. Some of the central factors highlighted by respondents as underpinning differentiation had clear power and inequality dimensions. These include political ties and corruption; community and family networks; and the capability to enforce one's own rights, for example to land, in lieu of access to impartial law enforcement and justice institutions. The article concludes with implications of these findings for supporting adaptive capacity and mainstreaming climate change into planning processes at different levels. In particular, interventions focused on assets may be less important than those directed at power relations, networks and the security dimensions of extreme poverty.

1. Introduction

Poverty and climate change vulnerability are widely regarded as being closely linked; however, this is based on a fairly generic understanding of vulnerability and adaptive capacity. There remains a scarcity of empirical research on how climate-related impacts affect livelihoods across and within groups of poor people on the ground (Tanner & Mitchell, 2008). The livelihoods of extremely poor people depend on similar climate-sensitive resources to those of less poor farmers, but in importantly distinct ways; however this has been the subject of relatively less climate change research to date. The research presented below (Coirolo, 2013) represents a critical gap, given that some regions with the highest concentrations of extreme poverty are also likely to experience the most severe impacts from climate change (Scott, 2008).

Poverty can be measured in various ways. The Bangladesh Bureau of Statistics (BBS) uses lower and upper poverty lines, which capture households whose total expenditures (consumption) are less than the amount needed to acquire a basic food basket of 2122 calories per person per day. The upper poverty line captures households whose expenditures are less than that needed to acquire both the food basket and a set of non-food items consumed by those close to the food poverty line. The lower line is a measure of extreme poverty, the upper line of moderate poverty. BBS also computes a poverty gap and squared poverty gap to measure the depth and severity of poverty, respectively (BBS, 2011b). Several other measures of poverty commonly used for cross-country comparisons include the share of the population living on less than US$2 and less than US$1.25 per day, calculated by the World Bank; and the Human Development Index and Millennium Development Goals, both of which include a broader set of non-income indicators.

For this research, participatory wealth-ranking activities (Chambers, 1997; Narayan, Patel, Schafft, Rademacher, & Koch-Schulte, 1999) were used to understand the socio-economic breakdown in the fieldwork communities, exploring in-depth what it means to be rich, or poor, in local terms, and to provide an initial identification of extremely poor households in both field sites. This approach allowed respondents to identify for themselves the socio-economic groups in their villages, and opened space for discussion of the more intangible, subjective elements of wealth and poverty that are largely missed by income and consumption-focused methods. The results of each participatory wealth ranking on poverty breakdown and extreme poverty criteria identified by respondents largely matched indicators used by non-governmental organizations (NGOs) and development agencies operating extreme poverty programmes in the local area, e.g. Gana Unnayan Kendra (GUK), BRAC and the UK Department for International Development (DFID).

This research departs from a 'starting point' approach (O'Brien, Eriksen, Nygaard, & Schjolden, 2007) in which vulnerability is seen as a context, shaped by socio-economic and political factors and processes that exist prior to the occurrence of a hazard, and that mediate exposure and sensitivity to it, as well as the ability of individuals and communities to cope with its effects. Vulnerability is thus underpinned by factors such as inequality in access to resources (Blaikie, Cannon, Davis, & Wisner, 1994; Paavola & Adger, 2006; Tschakert, 2007). This often results in a relative inability to prepare for and successfully cope with the effects of shocks and stresses, including climate change, among poor and socially excluded communities, and of particular individuals within these communities (e.g. children, women, disabled individuals).

Vulnerability to the impacts associated with climate change is therefore differentiated across groups and individuals. This underscores the importance not only of working at the level of groups but also of exploring the factors that shape differentiated vulnerability between individuals belonging to the same group. A major factor underlying differential vulnerability in the research presented here is structural inequality and asymmetries of power that, in turn, mean that certain groups and individuals are disproportionately exposed to impacts from climate events, their livelihoods are more climate dependent and sensitive, and they lack access to resources for coping successfully with shocks, let alone adapting to climate change. This means that they often feel the greatest impacts and take the longest to recover from shocks and stresses.

This paper presents findings from Coirolo (2013) on factors that underpin differential levels of vulnerability across two communities in Northwest Bangladesh, with a focus on extremely poor community members. The central aim is to contribute to an evolving body of empirical data on the role of intangible assets, factors and processes in shaping differential vulnerability to the impacts of climate-related shocks and stresses, and differential capacities to adapt to changing climatic conditions. Power-related factors are explored towards the end of understanding where gaps and needs exist with respect to building adaptive capacity among extremely poor people.

2. Description of research: theory, fieldwork location and methods

2.1. Conceptual framing

The livelihoods framework (DFID, 1999) was used as a conceptual model and guide to data collection for analysing vulnerability among respondents. The re-framed concepts of *resources* (rather than capitals) and *mediating factors*, i.e. the factors and processes that influence levels of climate-related vulnerability and adaptive capacity among extremely poor respondents, were added onto the livelihoods approach, forming core components of a mediating factors framework (MFF)[1] that was developed for this research (Figure 1).

The choice to employ the concepts of resources and mediating factors was made in order to achieve a truly respondent-led approach. This research uses the term *resources* rather than capitals to avoid the reliance on pre-defined categories of assets that may be less relevant in exploring the livelihoods of respondents in a given field site location. Instead, the central focus was on how respondents themselves defined and categorized the resources they consider important to their livelihoods and coping strategies.

This research also borrows elements of the resource profile framework (Lewis, 1993), which employs a broader and less 'taxonomic' livelihoods approach by using the term 'resources' to accommodate a wider range of livelihood components. This opens space for

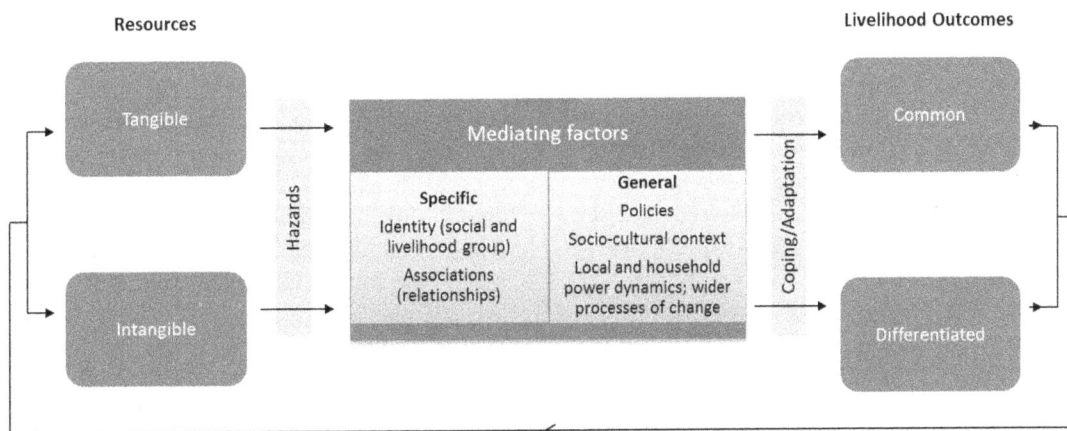

Figure 1. Mediating factors framework.
Note: Coastal or riverine islands formed by cycles of erosion and accretion of sediment.

consideration of cultural and spiritual factors, the role of power resources (Korpi, 1985), perceptions of risk, and community networks and relations, while paying greater attention to the variable, fluid nature of resource values and the relationships and activities that underpin this fluidity.

Mediating factors operate by mediating the ability of individuals and households to access and deploy tangible and intangible resources in coping with the impacts of hazards (shocks and stresses, climate-related or otherwise) on their livelihoods. In this way, mediating factors shape differential levels of vulnerability and adaptive capacity across the various sub-groups and individuals that make up the 'extreme poor'. These sub-groups include respondents engaged in different livelihood activities (mainly river-based versus agricultural) and social categories (i.e. men, women, children, elderly, disabled individuals), as well as differentiation that may exist between individuals within these groups. For instance, not all fishermen in a community are affected in the same way and to the same degree by a flood or storm, just as not all young children or women are equally able to cope with the effects of drought. Rather, additional patterns of differentiation emerge between individuals who pertain to the same livelihood or social group, and the mediating factors element of the framework is designed to capture these additional dimensions.

This is reflected in the MFF through use of 'specific' and 'general' mediating factors, with examples used to illustrate the types of factors that might mediate vulnerability and adaptive capacity at the individual level (e.g. identity-based characteristics, such as livelihood and social group, networks and relationships) as well as at the household and community levels (e.g. broader processes of change, local power dynamics, institutions) (Smit & Wandel, 2006).

2.2. *Fieldwork location*

Data collection was undertaken in two sites in Gaibandha District, Northwest Bangladesh. Meteorological data indicate that temperature and precipitation patterns have started to break with historical trends over recent decades, both across Bangladesh and in the Northwest region. There has been a related increase in the frequency and variability of the kinds of climate events the country has historically experienced (CCC, 2009; Habib, 2011; Islam, Ferdousi, Rahman, Ahsan, & Abdullah, 2008). Temperatures have increased by 5°C over the last 100 years in Bogra, the District neighbouring Gaibandha, where the nearest weather station is located (Islam, undated). Precipitation patterns have become increasingly erratic across the entire country, with significant increases recorded in the Northwest region (Shahid, 2010). The incidence and severity of extreme events, such flooding and severe drought, have increased in the Northwest region

over the second half of the twentieth century (Asada & Matsumoto, 2009; Rajib, Mortuza, Selmi, Ankur, & Rahman, 2008), as has erosion, due to 'climate change-induced intensifying rainfall pattern and unplanned interventions' (Uddin & Basak, 2012, p. 3). These changing patterns are projected to increase in the future as climate change accelerates, with particularly adverse consequences for poor communities, who tend to rely disproportionately on climate-sensitive livelihoods, are more exposed to climate impacts given tendency to live in relatively marginal areas, and have limited capacity to cope with climate-related impacts in the present, much less adapt to evolving climate conditions over longer timescales (Khan, Ali, Asaduzzaman, & Bhuyan, 2010; Tanner & Horn-Phathanothai, 2014).

Gaibandha, one of the country's poorest and most disaster-prone districts, is a striking example of these issues (World Bank, BBS, & WFP, 2009). The District's location at the confluence of the Brahmaputra, Jamuna and Teesta Rivers makes it particularly vulnerable to floods, drought and riverbank erosion, all of which severely constrain livelihoods (Ensor & Berger, 2009; Khan et al., 2010). Fieldwork was undertaken in two sites in Gaibandha, including Bariakari, a *char* island community, and Rajiapur, a community residing on a river embankment. Both sites are characterized by high rates of extreme poverty and food insecurity; high reliance on agriculture and, in particular, agricultural day labour; high biophysical and social vulnerability to natural hazards; poor performance across basic human development indicators; and low access to government services and NGO development programmes, relative to both the rest of the country and other *upazilas*[2] in Gaibandha (BBS, 2011a, 2011b; BBS & UNICEF, 2010; BBS & WFP, 2004; World Bank et al., 2009).

2.3. *Methods*

Fieldwork took place in three discrete and sequential phases, each corresponding to increasingly lower levels of analysis (community, household, individual). These phases comprised: (1) community-level climate vulnerability and capacity analysis (CVCA) (Dazé, Ambrose, & Ehrhart, 2009), a set of participatory methods for assessing climate change vulnerability and adaptation among communities based on activities that blend participatory rural appraisal/livelihood approaches from development with disaster risk reduction (DRR)-based tools; (2) household semi-structured interviews, conducted over several visits, with every household member willing to participate interviewed, and; (3) individual life history interviews.

For the first phase, cross-sections of the entire community participated as respondents, through focus group discussions including individuals from different socioeconomic groups and pertaining to various different livelihood groups. For the next two phases, across the two main fieldwork sites, a group of 41 households (163 individuals) matching the criteria for extremely poor households identified during participatory wealth-ranking activities (CVCA phase) were selected to make up the core respondent group. These individuals participated in the more in-depth household and life history interviews comprising phases two and three of data collection, respectively.

3. Explanation of research results

3.1. *Overview*

Research findings illustrate how differential vulnerability is shaped by various 'mediating factors', many of which relate to power and inequality in access to and control over resources, both tangible and intangible. Wider patterns of differentiation emerge at the *community* level, for instance between the core, extremely poor respondent group and non-extreme poor community members, and among different social and livelihood groups. Additional layers of differentiation in vulnerability and adaptive capacity also emerge across households and individuals classified as extremely poor, as well as between members pertaining to the same social or livelihood groups. These additional dimensions of differentiation were identified by respondents as being shaped by sets of 'mediating factors' that operate at the sub-group level. This underscores the importance of analysis at the sub-group level in relation to patterns of vulnerability and need when designing climate change adaptation policies and programmes and mainstreaming adaptation into broader planning.

In addition to patterns of differentiation, areas of commonality were found across extremely poor respondents, including a general lack of ownership and decision-making power over the resources upon which one's livelihood depends (e.g. agricultural work most often through day labour rather than farming one's own land); perceptions of risks and livelihood impacts from climate change; and certain coping strategies that are common across all households.[3]

The following section focuses specifically on the power-related mediating factors that were highlighted as drivers of differentiation among extremely poor respondents.

3.2. *Power-related mediating factors*

Within the extremely poor group, some of the main mediating factors highlighted by community members had clear power and inequality dimensions. These could be classified as underpinning different kinds of 'power resources' or 'attributes (capacities or means) of actors (individuals or collectivities), which enable them to reward or punish other actors' (Korpi, 1985, p. 33). In the case of this research, each of the following power-related mediating factors played a role in shaping levels of vulnerability to shocks and stresses and adaptive capacity across households and individuals: *political ties and corruption*; *community and family networks*; *capability and power to*

enforce rights in lieu of access to impartial law enforcement and justice institutions.

3.2.1. *Political ties and corruption*

Those individuals who had relationships with locally powerful men or members of local government tended to receive preferential treatment in accessing official government assistance and relief; however, these were mainly households that were slightly better off (poor rather than extremely poor). Most extremely poor respondents faced barriers to accessing government assistance due to corruption. Allegations of needing to bribe government officials to access one's fair share of benefits, social safety nets (SSNs) and larger amounts of flood relief are common among extremely poor community members:

> Ashik: "They do not give me OAP (Old Age Pension), VGD (Vulnerable Group Development card), nothing. Other people ask me: 'you are poor, you are blind, you have no ability to work, why they don't give you card?' Member, chairman (of local government) don't give me any help. I have no money. *Saree*,[4] OAP, these type of help come from the government to the member and chairman, but we do not receive any of these help, the people who can pay, they receive these."

In both field sites rates of SSN access among respondents are reportedly decreasing: in Rajiapur, 27% of extremely poor respondents had received some SSN five years prior to fieldwork, as opposed to 11.5% at the time of fieldwork; in Bariakari, 60% of respondent households previously received an SSN, down to 30% at the time of fieldwork. A clear pattern emerged in the data whereby in both sites, access to government SSNs and relief (as well as some NGO programmes) is mediated, albeit in different ways, by a 'culture of corruption' (Marshall et al., 2010, p. 17), underpinned by political connections (though these are few among the respondent group) and bribery. While the expansion of SSNs has played an important role in reducing rural poverty in Bangladesh (BBS, 2011b), the trend for respondents in both research sites has been one of decreasing access over the last five years.

Relatively more extremely poor respondents were part of local NGO programmes than in the past. However, in addition to paying for access to SSNs, access to some NGO and government programmes for extremely poor respondents is controlled in other ways, for instance through locally powerful men, as in the case of 'middlemen'. These individuals use their political connections for personal gains by acting as intermediaries between NGOs/donors, local governments and communities, demanding bribes for including names on a list of suggested beneficiaries, leaving off the names of those who cannot pay.

The benefits of accessing both government programmes and NGO interventions would enhance adaptive capacity among respondents, as they aim to address poverty and vulnerability through, for instance, provision of basic services (health, education), access to basic amenities (water, sanitation) and transfer of income-generating assets. In terms of extreme poverty–climate change vulnerability links, a central implication of this research is that climate-related shocks and stresses may become an increasingly important driver of poverty for extremely poor people, to the extent that weather patterns and climate extremes continue to shift, and respondents continue to face barriers to accessing the resources necessary for implementing successful coping strategies. With respect to NGOs and local government, strategies to address the clear gap identified by various respondents – namely their inability to hold duty bearers to account in terms of acknowledging and upholding the rights of extremely poor individuals who are unable to pay bribes and lack the political connections to access benefits and support for coping with shocks and stresses – emerge as central for effective coping and mainstreamed adaptation planning and implementation that benefits the most vulnerable.

3.2.2. *Community and family networks*

Unofficial safety nets operating at the community level emerged as a critical resource for coping with shocks for many respondents. The *dhar koroj*[5] system operates throughout the year, during which it is customary among neighbours and fellow community members to lend and borrow small, everyday items, like salt and chilli for cooking, paddy or small sums of money. Respondents report falling back on this system during times of hardship, for instance during lack of work due to illness, but this is based on relationships that were built over some time, and therefore exists primarily among those with some degree of familiarity. This underpins the importance of building up social networks with other community members as a preventive coping strategy (Wisner, Blaikie, Cannon, & Davis, 2003); calling on friends and neighbours to help requires first building relationships with them.

Yaron: "Those who have affection and love with us in this place, we go to them during floods, to those who know us. Maybe they say to us that they don't have money right now, but here is rice, take that … at first when I arrived I did not know the people here, my neighbours, nobody would give to me then. But now we have very good relations, and they help me. I also give. If I don't give to them when they need, then when I will need and go to them, they will not give me."

However, there are also limits to community safety nets. For instance, they provide protection mainly from the effects of individual or idiosyncratic shocks, since they can easily become overwhelmed in the case of widespread or covariate shocks (Dercon, 2010) during which times family networks extended over larger geographical areas

became essential for respondents. This reflects literature on the nature of informal social and family networks as a potential element of adaptive capacity: 'for a network to enhance resilience, different nodes of the network must rely on different resources, or the resource must be heterogeneous in space or time' (Marshall et al., 2010, p. 15). While community *dhar koroj* is relied upon primarily when respondents experience idiosyncratic shocks, family support networks become critical for coping during times of covariate shock, with family members located in other districts that are unaffected by a given flood, drought or other shock, more able to provide assistance than fellow community members.

Mohiron: "I don't ask for food during flood. I ask in other times. During flood time everyone take shelter in the road and some people go to their relatives house, in other village … sometimes my daughter sends me packages with food, *saree*. She lives in Dhaka … Everybody here is busy with themselves, so nobody helps during the flood."

Another limitation of the community safety net is that certain individuals are excluded, including those who had recently moved into the local area and had not yet built up relationships with fellow community members, as well as the poorest of the poor, given their inability to reciprocate exchange of resources. While interest is not charged on small amounts of money borrowed between neighbours, it is expected that the money will be re-paid quickly, usually within days. In relation to exchanges of food, there is an expectation of repayment attached to lending certain kinds of food, e.g. uncooked rice; cooked food, on the other hand, is considered a gift. However, there is an expectation of reciprocity at some point in the future generally attached to gifts of food.

Zahir: "For money we have to return after borrow. But for food you sometimes don't have to return. Suppose I don't have rice then I go to someone and I ask that 'give me rice to eat,' then they give. In that case it is OK if I don't give it back. But if you take paddy or *taka*[6] then you must give back … if I ask anyone to give me 10 *taka hawlat* I must return it after 2 or 3 days. It is called *dhar koroj*."

For these reasons, the respondent households classified as 'poorest of the poor' with no ability to reciprocate in the future are generally not included in these unofficial community safety nets. For these households and individuals, the only option for accessing resources to cope with shocks when they occur is to beg and/or try to borrow from local money lenders:

Trina: "I want to take *taka hawlat* when I have no food. People don't give *hawlat* for long time, people want that you return quickly, but I am a beggar, I have no way to do this, so people don't want to give me *hawlat*. I have to repay *hawlat* according to date, and I have no work so how can I repay like this? This is the thing."

Collective decision-making is another feature of family and community networks that becomes a resource when coping with climate-related hazards. Respondents who are forced to relocate due to erosion tend to come together to make decisions about where to move next, and often do so in a group. Collective relocation represents perhaps the only coherent coping response available to erosion-affected *char* communities more generally in Bangladesh (Hutton & Haque, 2004). The findings here relate to literature on the role of community-level determinants of adaptive capacity, such as the capacity to re-organize, or 'effectively respond to disturbances in order to plan for disturbance' (Marshall et al., 2010, p. 16). While most of the erosion-induced relocation occurring among respondents is reactive, it nonetheless provides evidence of groups of respondents regularly carrying out the fairly major undertakings of re-establishing entire communities on new *chars*, often towards the end of retaining unofficial community safety nets, in a coordinated way in response to climate-related shocks, with little external support. This suggests that the potential exists for forward-planning, successful autonomous adaptation even among extremely poor communities.

> Prianka: "Say our *char* started to erode. Then we must find another *char*. We talk with the other villagers and tell them 'let's go there. All together we can live there.' Then together we take down the houses, separate the walls, and all move together ... then if I have no food I can easily borrow from them. If we move alone, we won't know those people in the new place, why would they help us?"

On the other hand, prior experience does not necessarily prepare communities for potentially increasing rates of erosion, and other shocks and stresses, as well as changes in timing and therefore predictability with which these events may occur in the future under accelerating climate change.

> Zubayer: "I am river-eroded people ... Because of erosion, I am moving from one *char* to another *char*. Maybe 20 times already in my life. This is my whole life. And it now is going worse than before when I was a child."

The central role of community and family networks in underlying response to climate-related shocks and stresses, while beneficial, was a testament to the low levels of respondents' power and capacity to hold government to account for delivering programmes and resources to which respondents were in principle entitled. While expansion of SSNs and agricultural research and extension have played an important role in reducing rural poverty generally in Bangladesh (BBS, 2011b), access to these services was still quite limited among the extremely poor respondent group.

3.2.3. *Capability and power to enforce rights*

The capability to enforce rights to access resources such as land emerged as a particularly important factor underpinning vulnerability and adaptive capacity. This was illustrated in the frequency of violent conflict over land as well as the common practice of bringing false court cases against individuals as a means of ousting them from their land. The importance of physical strength and capability was also illustrated through the inability of older widows to maintain landholdings after the death of their husbands, after which they were often forcibly removed. This is not only a problem for widows, but was common even in households with male members, as in the case of Mamun and his family:

> "It's my father's land. I was in Rangpur when my father died, at that time he divided the land for us. We were six brothers, three of them now are dead, and after dividing the land of my father's I get two *bigha*[7] land ... Actually, my nephew, son of one brother, he use the land and I am not able to use this land at all. When I came from Rangpur after death of my father, at that time my brother and their sons were using the land, but when I came here I said 'give me my land,' but they did not want to give. So the fight is start then ... I try to cultivate this land anyhow but then they come and beat me. Because I am old and alone here and they live with their family. They come more powerful ... This year my eldest son was here for two months, he lives in Dhaka. My son and wife's nephew and the husband of my niece, we all go together to the land and say 'you are using my land and don't give any money for using my land, we will cultivate this land.' My brother was so angry he gave a false rape case against me, but the chairman dismiss the case because there was no witness ... still though I don't use the land now. My son is in Dhaka now and I don't manage alone to use the land."

This type of land dispute between community and family members was found to be common, as are conflicts resulting from influxes of new households into the local area looking for new land following erosion. Extremely poor respondents face a set of circumstances in these situations that are challenging to overcome: they lack the financial resources and/or political connections to disentangle easily from false court cases, to bribe police officers or purchase new land; they are also often in poor health and malnourished, making the prospect of physical altercations risky, especially given high reliance on physical labour for income.

This reflects the concept of power resources (Korpi, 1985). Korpi (1985) identifies the longstanding division in the tradition of studying power between approaches that define power with reference to its expression or situations in which it is exercised, and those that instead view power as an 'ability or capacity' (p. 33), or resource. The latter approach opens space for analysis of the role of power resources even when unused. In the case of this research, the physical need to protect one's right to resources, and the actual and perceived lack of power resources (physical and financial) on behalf of extremely poor respondents, leaves them trapped in a position in which they are viewed by others as powerless to resist physical coercion, and therefore often fall victim to stronger and more powerful individuals. In lieu of impartial justice institutions, this leaves them with no means of acquiring valuable resources or protecting ones they have. In the context of increased rates of erosion under a changing climate, these findings imply an urgent need to address the gap in access to official support for extremely poor, erosion-affected communities.

4. Conclusions

The research presented above illustrates the central role of power-related factors in shaping differential vulnerability across extremely poor community members in two fieldwork sites in Northwest Bangladesh. This highlights the importance of exploring the non-climatic, structural and socio-political drivers of vulnerability in a given setting, and of taking a mediating factors approach that places the perceptions and views of respondents – or intended programme beneficiaries – centre stage.

While differentiation has been the central focus of this paper, one commonality that defines most if not all of the extremely poor respondents is lack of control over the resources and activities upon which their livelihood security depends. The room for manoeuvre and scope for decision-making of the extreme poor are far more circumscribed than is the case for the less or non-poor. The extent to which resources critical to their survival and well-being are climate-sensitive or not is frequently determined by decisions of others.

This suggests the need for a shift in strategic emphasis of policies and programmes aimed at mainstreaming climate change adaptation. Central messages of this research would be the critical importance of addressing the underlying root causes and drivers of *differentiated* vulnerability for extremely poor and excluded groups. For such groups, interventions focused on assets may be less relevant than those directed at power relations, networks and the security dimensions of extreme poverty. Reserving space and time for beneficiary inputs into the design, implementation and subsequently monitoring and evaluation of activities emerge as particularly important.

Strengthening the voice of extremely poor and excluded people within decision-making processes will require greater emphasis on 'the processes through which communities are able to make changes to their lives and livelihoods in response to emerging environmental change' (Ensor, 2011, p. 2). This is echoed by the view that what is new and important about an adaptive capacity perspective relative to a focus on discrete adaptation interventions is exploring 'what a system *does* that enables it to adapt, rather than what a system *has*' (WRI, 2009, cited in Levine, Ludi, & Jones, 2011, p. 5). Areas where policy could therefore support local adaptive capacity include power sharing; knowledge and information; and

experimentation and testing. This kind of approach would require a 'reorientation of development thinking' to focus on *processes* – through which adaptation results would emerge – rather than the more traditional emphasis on interventions and outcomes (Ensor, 2011, p. 33).

An important difference with respect to current poverty/livelihood/development strategies may be less about the activities undertaken in pursuit of climate change adaptation per se, and rather more about adjusting timescales, sequencing and information on which activities are planned, and flexibility in terms of responsiveness to local conditions and the evolving needs of beneficiaries. In designing or adjusting poverty and livelihood programmes for extremely poor communities in climate-sensitive areas, a potential indicator for an 'adapted' community could be flexibility or ability to change livelihood activity easily, in response to and in preparation for weather patterns and climatic hazards in the present as well as more gradual transformation towards climate-resilient livelihoods over the longer term. A strategic focus on flexibility may well affect the approach to timescales and frequency of existing interventions (e.g. flood-coping support four times during flood season rather than once or twice, if needed) and recognition of emerging needs that arise due to progressively shortening periods of time to recover between floods.

In terms of mainstreaming climate change into planning processes, the need to address underlying drivers of vulnerability would entail a shift beyond short-term, reactive coping to include support for adaptive transformation of livelihoods in line with changing climate conditions over longer timescales. This could involve extension of current safety net programmes to address the medium-term effects of disasters, taking account of differential impacts across social and livelihood groups (Coirolo, Commins, Haque, & Pierce, 2013). It would also require more coherent integration of policy areas that aim to reduce vulnerability, including climate change adaptation, DRR and social protection, into an adaptive social protection framework (Arnall, Oswald, Davies, Mitchell, & Coirolo, 2010; Davies et al., 2013; Davies, Guenther, Leavy, Mitchell, & Tanner, 2008). This would engender a more integrated and holistic approach to vulnerability over different timescales, including policies aimed at protecting vulnerable groups, preventing losses from shock events and improving security, such as through access to basic health and education services, and would extend to interventions that address differential power and access to rights through empowerment of excluded and marginalized groups. The power-related factors that emerge as particularly important for the extreme poor respondents in the case of this research include establishment of impartial police and justice services, land reform and distribution for communities displaced by events like erosion, and discouraging political networks that tolerate corruption and reinforce exclusion of extremely poor people.

Notes

1. Only the selected elements of the MFF that relate to the topic of this article are described in detail here. For further elaboration see Coirolo (2013).
2. Administrative unit below Sub-Division, also known as *Thana*.
3. For a full elaboration of research results, see Coirolo (2013).
4. Garment worn by females.
5. Unofficial community safety net, also known as *hawlat*.
6. National currency of Bangladesh.
7. A unit of land measurement equivalent to 0.2 acres.

References

Arnall, A., Oswald, K., Davies, M., Mitchell, T., & Coirolo, C. (2010). *Adaptive social protection: Mapping the evidence and policy context in the agriculture sector in South Asia*. Working Paper 345. Institute of Development Studies, Brighton.

Asada, H., & Matsumoto, J. (2009). Effects of rainfall variation on rice production in the Ganges-Brahmaputra basin. *Climate Research, 38*, 249–260.

Bangladesh Bureau of Statistics (BBS). (2011a). *Bangladesh population and housing census 2011*. Dhaka: Author.

Bangladesh Bureau of Statistics (BBS). (2011b). *Preliminary report on household income and expenditure survey-2010*. Dhaka: Author, Statistics Division, Ministry of Planning.

Bangladesh Bureau of Statistics (BBS), & UNICEF. (2010). *Key findings of the Bangladesh multiple indicator cluster survey 2009: Monitoring the situation of women and children*. Dhaka: Author.

Bangladesh Bureau of Statistics (BBS), & World Food Programme (WFP). (2004). *Local estimation of poverty and malnutrition in Bangladesh*. Dhaka: Ministry of Planning, Planning Division, Statistics Wing, Author.

Blaikie, P., Cannon, T., Davis, I., & Wisner, B. (1994). *At risk: Natural hazards, people's vulnerability, and disasters*. London: Routledge.

Chambers, R. (1997). *Whose reality counts? Putting the first last*. London: ITDG.

Climate Change Cell (CCC). (2009). *Characterizing long-term changes of Bangladesh climate in context of agriculture and irrigation*. Dhaka: Author, DoE, MoEF; Component 4b, CDMP, MoFDM.

Coirolo, C. (2013). *Climate change and livelihoods in Northwest Bangladesh: Vulnerability and adaptation among extremely poor people*. Brighton: Institute of Development Studies (IDS), University of Sussex.

Coirolo, C., Commins, S., Haque, I., & Pierce, G. (2013). Climate change and social protection in Bangladesh: Are existing programs able to address the impacts of climate change? *Development Policy Review, 31*(s2), 74–90.

Davies, M., Béné, C., Arnall, A., Tanner, T., Newsham, A., & Coirolo, C. (2013). Promoting resilient livelihoods through adaptive social protection: Lessons from 124 programmes in South Asia. *Development Policy Review, 31*(1), 27–58.

Davies, M., Guenther, B., Leavy, J., Mitchell, T., & Tanner, T. (2008). Adaptive social protection: Synergies for poverty reduction. *IDS Bulletin, 39*(4), 105–113.

Dazé, A., Ambrose, K., & Ehrhart, C. (2009). *Climate vulnerability and capacity analysis handbook* (1st ed.). CARE International.

Dercon, S. (2010). Risk, poverty, and human development: What do we know, what do we need to know?. In R. Fuentes-Nieva & P. A. Seck (Eds.), *Risk, shocks, and human development: On the brink* (pp. 15–39). New York: Palgrave Macmillan.

DFID. (1999). *Overview. Key sheets for sustainable livelihoods*. Department for International Development Natural Resources Policy and Advisory Department. Retrieved December 9, 2008, from http://www.oneworld.org/odi/keysheets/

Ensor, J. (2011). *Uncertain futures: Adapting development to a changing climate*. Warwickshire: Practical Action Publishing.

Ensor, J., & Berger, R. (2009). *Understanding climate change adaptation: Lessons from community-based approaches*. Warwickshire: Practical Action Publishing.

Habib, A. (2011). *Climate change: Bangladesh perspective*. Dhaka: Bangladesh Meteorological Department.

Hutton, D., & Haque, C.E. (2004). Human vulnerability, dislocation and resettlement: Adaptation processes of river-bank erosion-induced displaces in Bangladesh. *Disasters, 28*(1), 41–62.

Islam, A.K.M.S. (undated). *Analyzing changes of temperature over Bangladesh due to global warming using historic data*. Dhaka: Bangladesh University of Engineering and Technology (BUET).

Islam, M.N., Ferdousi, N., Rahman, M.A., Ahsan, M.N., & Abdullah, S.M.A. (2008). *Understanding the rainfall climatology and detection of extreme weather events in SAARC region: Part I – Bangladesh. A collaborative research project between Synoptic Division of SMRC and HyARC of Nagoya University*. Dhaka: SAARC Meteorological Research Centre (SMRC).

Khan, I.A., Ali, Z., Asaduzzaman, M., & Bhuyan, M.H.R. (2010). *The social dimensions of adaptation to climate change in Bangladesh*. Washington, DC: World Bank.

Korpi, W. (1985, Autumn). Power resources approach vs. action and conflict: On causal and intentional explanations. *Sociological Theory, 3*(2), 31–45.

Levine, S., Ludi, E., & Jones, L. (2011). *Rethinking support for adaptive capacity to climate change: The role of development interventions. A report for the Africa climate change resilience alliance*. London: ODI.

Lewis, D.J. (1993). Going it alone: Female-headed households, rights and resources in rural Bangladesh. *European Journal of Development Studies, 5*(2), 23–42.

Marshall, N.A., Marshall, P.A., Tamelander, J., Obura, D., Malleret-King, D., & Cinner, J.E. (2010). *A framework for social adaptation to climate change: Sustaining tropical coastal communities and industries*. Gland: IUCN.

Narayan, D., Patel, R., Schafft, K., Rademacher, A., & Koch-Schulte, S. (1999). *Can anyone hear us? Voices from 47 countries*. Poverty group, PREM. Washington, DC: World Bank.

O'Brien, K., Eriksen, S., Nygaard, L.P., & Schjolden, A. (2007). Why different interpretations of vulnerability matter in climate change discourses. *Climate Policy, 7*, 73–88.

Paavola, J., & Adger, W.N. (2006). Fair adaptation to climate change. *Ecological Economics, 56*, 594–609.

Rajib, M.A., Mortuza, M.R., Selmi, S., Ankur, A.K., & Rahman, M.M. (2008). *Spatial distribution of drought in the Northwestern part of Bangladesh*. Dhaka: Bangladesh University of Engineering and Technology (BUET).

Scott, L. (2008). *Climate variability and climate change: Implications for chronic poverty*. Working Paper No. 108. Chronic Poverty Research Centre (CPRC).

Shahid, S. (2010). Recent trends in the climate of Bangladesh. *Climate Research, 42*(3), 185–193.

Smit, B., & Wandel, J. (2006). Adaptation, adaptive capacity and vulnerability. *Global Environmental Change, 16*, 282–292.

Tanner, T., & Horn-Phathanothai, L. (2014). *Climate change and development*. New York: Routledge.

Tanner, T., & Mitchell, T. (2008). Entrenchment or enhancement: Could climate change adaptation help to reduce chronic poverty?. *IDS Bulletin, 39*(4), 6–16.

Tschakert, P. (2007). Views from the vulnerable: Understanding climatic and other stressors in the Sahel. *Global Environmental Change, 17*(3–4), 381–396.

Uddin, A.F.M.A., & Basak, J.K. (2012). *Effects of riverbank erosion on livelihoods*. Dhaka: Unnayan Onneshan.

Wisner, B., Blaikie, P., Cannon, T., & Davis, I. (2003). *At risk: Natural hazards, people's vulnerability and disasters* (2nd ed.). London: Routledge.

World Bank, BBS, & WFP. (2009). *Updating poverty maps of Bangladesh*. Dhaka: Author. Retrieved June 13, 2009, from http://home.wfp.org/stellent/groups/public/documents/liaison_offices/wfp201528.pdf.

REVIEW ARTICLE

Moving towards inclusive urban adaptation: approaches to integrating community-based adaptation to climate change at city and national scale

Diane Archer[a], Florencia Almansi[b], Michael DiGregorio[c], Debra Roberts[d], Divya Sharma[e] and Denia Syam[f]

[a]International Institute for Environment and Development, London, UK; [b]Instituto Internacional de Medio Ambiente y Desarrollo-América Latina, Argentina; [c]The Asia Foundation, Hanoi, Vietnam; [d]Environmental Planning and Climate Protection Department, Durban, South Africa; [e]Sustainable Habitat Division, The Energy and Resources Institute (TERI), New Delhi, India; [f]Mercy Corps (Indonesia Country Office), Indonesia.

Adaptation to climate change in urban areas presents a complex challenge. Consequently, approaches to urban adaptation should be both multilevel and multidimensional. Community-based adaptation (CBA) presents an opportunity for local-level participation in framing adaptation planning and activities, with wider transformative potential for urban governance. This paper presents five case studies from cities in the Global South which offer insights into the different scales at which CBA can be mainstreamed in urban contexts, and the various ways in which this is happening. These examples demonstrate five emerging opportunities for mainstreaming urban CBA, which include using CBA as part of a wider package of approaches; seizing processes of institutional reform as an opportunity to integrate community perspectives; institutionalizing new actors and approaches as a mechanism for scaling up multi-stakeholder approaches; ensuring top-down planning approaches are connected to local dynamics; and using participatory research to facilitate local communities in shaping planning processes. The cases also demonstrate that while obstacles to mainstreaming in urban contexts remain, some lessons in addressing these challenges have emerged, and CBA should, therefore, be a part of the toolbox of local and national urban adaptation policy frameworks.

Introduction

In a world in which the majority of the population lives in urban centres, the importance of addressing urban vulnerability to climate change cannot be underestimated. Urban areas represent complex systems and concentrations of risk, and both rapid and slow onset disasters can be seen as a result of failures in urban governance (Dodman & Satterthwaite, 2008), with the potential to be aggravated if the impacts of climate change are not adequately planned for. One billion people across the world live in informal settlements, lacking infrastructure, basic services, and secure housing, and are, therefore, particularly exposed to the impacts of climate change (Satterthwaite & Mitlin, 2014). Adequate building regulations, universal provision of infrastructure and services, and emergency preparedness may frequently be beyond the capacity or power of local authorities, while appropriate legal, financial, and insurance systems may be lacking (Dodman & Satterthwaite, 2008; Satterthwaite, Huq, Reid, Pelling, & Romero Lankao,

2007). There is, therefore, also a role for civil society to prepare for climate change, both through local non-governmental organizations (NGOs) and local communities, such as through community-based adaptation (CBA), which integrates governance approaches and tools for participatory planning.

Urban CBA can be seen as a response to rapid urbanization and competing pressures for scarce urban resources which may put communities at increased risk of climate change impacts through displacement effects (Soltesova, Brown, Dayal, & Dodman, 2014). As urban climate change adaptation remains a relatively new area of action for cities, and possible adaptation actions need to be context-specific, there are currently no standards or norms for planning and adaptation at the city scale, and different cities have adopted varying approaches to planning and implementing adaptation actions (Anguelovski & Carmin, 2011). Where community-level actions can be mainstreamed into, and supported by, city-level planning mechanisms, this creates the potential for more effective risk reduction whilst building capacity, devolving authority to the community level, enhancing governance and accountability.

CBA refers to the participatory identification and implementation of community-based development activities that strengthen the capacity of local people to adapt to climate change, and building on communities' expressed needs and perceptions to address local development concerns which underlie vulnerability (Ayers & Forsyth, 2009; Reid et al., 2009). Recognizing that adaptation is embedded within an institutional system which may have particular goals (Cannon & Muller-Mahn, 2010), community-led approaches to climate change adaptation may, therefore, either need to challenge both the existing institutional system and its development goals or be better integrated with it. Because CBA specifically seeks to engage with poor and more vulnerable people (Forsyth, 2013), it presents an opportunity to address the social, economic, and political drivers of vulnerability as part of broader development processes. Recognizing the challenges of defining 'communities' in the urban context, for the purposes of this paper they are regarded as residents of a particular area who are vulnerable to similar climate impacts. These residents may be socially and financially heterogeneous, but may be grouped through local administrative boundaries, issue-based local organizations, common interests, values, and activities (Twigg, 2007).

While the need to mainstream climate change adaptation into development planning and decision-making processes has been recognized, the resulting guidelines and tools developed by international agencies and NGOs may have contributed to a lack of clarity regarding mainstreaming (Olhoff & Schaer, 2010). Mainstreaming here refers to the integration of climate resilience considerations into development planning objectives and processes from national to local scales (Pervin et al., 2013), recognizing that a transformative approach which positively impacts the development agenda, is preferable to a merely 'additional' approach (Jahan, 1995, in Pervin et al., 2013). This paper explores the opportunities, benefits, and challenges to mainstreaming CBA into urban climate governance, 'in which public, private, and civil society actors and institutions articulate climate goals, exercise influence and authority, and manage urban climate planning and implementation processes' (Anguelovski & Carmin, 2011, p. 169).

Planning is in many cases a top-down process: from national structures to the city to the community level, with limited opportunities for integrating local-level actors in the process. When considering climate change, planning may include vulnerability and risk assessment processes, including community-level assessments – but significant gaps remain in inclusive approaches to urban adaptation which make room for and learn from community-level knowledge and adaptation actions. Recognizing that climate change is a challenge which cannot be addressed solely by a single organization or governance

institution, there is, therefore, a need for 'multi-level or multiscale governance' (Leck & Simon, 2012). This involves agents and institutions from both the government and non-government sectors, including community groups, through inclusion of community voices in defining the problem and finding solutions, and mainstreaming CBA approaches into planning and processes.

This paper draws on case studies of approaches to urban adaptation across five cities in the global South, which provide examples of cities currently implementing resilience-building initiatives through a variety of means and drivers. These cases were presented in the urban plenary session at the 2013 Community Based Adaptation 7 conference on 'Mainstreaming CBA' in Dhaka, Bangladesh, representing some key emerging issues across geographies. The cases consider how urban communities can and have mobilized to pursue adaptation strategies, and how local governments and other actors have incorporated these experiences, priorities, and capacities, in ways that mainstreams urban CBA. Presented from the perspective of local government, civil society, and academia, these examples, while being context-specific, represent emerging opportunities for mainstreaming CBA in urban areas, as follows:

- CBA is part of a wider package of approaches, with the appropriate institutional framing, as in Durban, South Africa;
- CBA offers an entry-point for aligning top-down approaches with local priorities, as in Guwahati, India;
- Processes of institutional reform are opportunities to integrate community perspectives, such as in Uruguay;
- The institutionalization of new actors, approaches, and funding mechanisms is facilitating the scaling up of multi-stakeholder approaches, as is happening in Indonesia;
- Participatory research plays a role in allowing local communities to shape planning processes (Forsyth, 2013), like in Quy Nhon, Vietnam.

While exploring the lessons emerging from current experiences, the paper will also demonstrate the relevance of CBA in building urban resilience to climate change, as well as the benefits of mainstreaming CBA into city- and national-level adaptation planning. The case studies, while limited in number, allow an exposition of various mechanisms for integrating community-level actions into city-level planning processes, as well as challenges which may impede integration, and how they have been or might be overcome. By exploring the scope for mainstreaming CBA at different scales and through a variety of drivers, the emerging lessons provide insights into approaches to building urban resilience which integrate community-level practices into planning at both the city and national levels, with subsequent potential for reshaping economic, social, and political institutions.

CBA within urban climate governance

While there is significant interest in CBA, existing studies have had a limited urban focus and highlight the gaps between local communities and public institutions. Case studies from Bangladesh offer insight into the potential of CBA in urban low-income settlements. Jabeen, Johnson, and Allen (2010) draw on the coping strategies of the urban poor in Korail slum in Dhaka to demonstrate that these strategies can be both preventative and impact-minimizing, but highlight the need to integrate local knowledge in pro-poor adaptation planning, while at the same time promoting democratic and accountable local governance structures to raise awareness of risks and ensure their integration into land use plans. They also point out that many of the most effective adaptation strategies require local government intervention to be implemented at scale, such as drainage systems. Roy, Hulme, and Jahan (2013) examine differences in adaptive behaviours of squatters who 'own' their homes and those who rent dwellings, concluding that governmental and non-governmental organizations must take into account land tenure status when deriving and implementing adaptation plans in urban low-income settlements. Both studies demonstrate numerous initiatives being undertaken at household and community levels which can be regarded as self-help measures, and which, with wider institutional support and provision of basic services, have the potential to adapt urban communities to climate change.

The case for mainstreaming CBA in urban governance

Urban climate governance remains an emerging policy domain. Carmin, Anguelovski, and Roberts (2012) argue that newer domains like climate change adaptation, endogenous goals, and objectives are more likely to drive adaptation actions at the city level, rather than exogenous factors like regulation or fundraising needs. In Durban and Quito, the endogenous drivers were specifically the cities realizing their vulnerability to climate change impacts, the efforts of champions pushing the agenda forward (a single champion in Durban, as illustrated below; elected officials in Quito), and adaptation being seen as a means to secure the cities' socio-economic development. Concurrently, 'the lack of resources, capacity, and best practices available to support climate action may be promoting innovation, attention to the most crucial needs and subpopulations, and the advancement of policies and initiatives that are grounded in local cultures and realities' (Anguelovski & Carmin, 2011, p. 173), which may be successful in the absence of national government policy.

New approaches to urban governance and institutions to promote resilience are one of the key focus areas of urban resilience literatures, including bundling resilience with broader development efforts in order to contribute to long-term sustainability (Leichenko, 2011). The IPCC (2007) defines resilience as 'the ability of a social or ecological system to absorb disturbances while retaining the same basic structure and ways of functioning, the capacity of self-organization, and the capacity to adapt to stress and change.' In urban contexts, resilience can be framed as the interaction between agents (from households to social organizations); systems of physical infrastructure and ecosystems; and formal and informal institutions, which govern relations between agents and systems (Tyler & Moench, 2012). This framework can facilitate urban adaptation planning by bringing together considerations of physical, ecological, social, and institutional factors.

However, the concept of resilience has been challenged as being insufficiently transformative, for promoting 'bouncing back' rather than 'bouncing forward' (Roberts et al., 2012), and not recognizing that often the current state of many institutional systems is the cause of problems. Pelling (2011) argues that adaptation presents an opportunity to move from the status quo of resilience towards transformation, that is, changes to the balance of power in society, through social reform and improved governance. While mainstreaming of CBA by itself will not automatically be transformative, this paper argues that approaches to urban resilience which give more room to community voices can reshape the definition of climate-related problems and hence solutions to them, in such a way that urban governance becomes more inclusive, transparent, and accountable.

Approaches to mainstreaming urban CBA

Mainstreaming of CBA cannot be considered outside of the wider policy-making context of adaptation, and the link between national- and local-level adaptation planning and the local communities on the ground. Ayers (2011) examines the potential for participatory, deliberative adaptation policy-making in the context of developing the National Adaptation Plan of Action (NAPA) in Bangladesh. She argues that for adaptation governance, it is necessary to move beyond simply creating institutional spaces for participation, towards 'deliberative governance, which should provide arenas for "risk-based" and "vulnerability based" adaptation discourses to come together and be resolved' (Ayers, 2011, p. 68). She found that adaptation priorities identified in the NAPA were based on expert-driven physical, impacts-based framings of risk, whereas local adaptation priorities were based on addressing development factors which made people vulnerable to climate impacts in the first place (Ayers, 2011, p. 81). Ayers recommends that deliberative adaptation governance should engage local institutional structures in adaptation planning

COMMUNITY-BASED ADAPTATION

from the beginning of the process, allowing CBA to be a starting point for adaptation planning, rather than the endpoint – while these findings are based on rural cases, they have resonance for urban areas.

For meaningful and deliberative governance, it is important to understand the role and dynamics of institutions functioning at different levels which can shape adaptation responses, whether they are public, private, or civil society institutions. Agrawal, Kononen, and Perrin (2009, p. 10) suggest that this shaping of adaptive capacity is determined by these institutions in three ways: they structure the nature of impacts and vulnerability to climate change through governance and communications; they create the incentive framework in which individual and collective actions take place; and they control access to resources and information which shape adaptation practices locally. Applied to an urban context, this relates to household provision of basic services and infrastructure by the state, access to financial institutions, and insurance mechanisms, which will shape collective and individual responses by local households (Satterthwaite et al., 2007). Dodman and Mitlin (2013) highlight the limitations of CBA in this regard, given the need for careful consideration of the potential of community-led projects to influence political and institutional structures, in order to contest the role of the state and necessary redistribution of resources. Examples from African and Asian urban poor federations demonstrate the potential for collective approaches to challenge underlying political and social structures for transformative effect (Dodman & Mitlin, 2013).

Planning plays an important role in adaptation given its forward-looking nature and its institutional basis within government (Hurlimann & March, 2012) – however, market mechanisms and politics mean that planning cannot be considered outside of power relations. Some cities may lack the flexibility to plan for uncertain futures, while adaptation plans may be motivated or influenced by external strategies or funders (Johnson & Breil, 2012). Nevertheless, effective and balanced planning for climate change adaptation can support action across spatial, temporal, and governance scales, in a way which maintains the public good while balancing competing interests (Hurlimann & March, 2012). However, planning as an effective and responsive adaptive mechanism requires sufficient capacity from planners and government institutions, as well as an informed public with the mechanisms to participate effectively, to ensure goals are not maladaptive. Few, Brown, and Tompkins (2007) highlight difficulties of planning for adaptation, where balancing immediate problems against a strategic, longer term perspective can make consensus difficult. Stakeholders may be selectively included or excluded, given the difficulties of defining the 'community' (Few et al., 2007).

While inclusive, deliberative approaches to urban climate governance are desirable, in practice their feasibility is constrained by capacity gaps, power relations,

and politics, which may limit the transformative potential of such an approach in an urban context. However, inclusive, multi-stakeholder approaches to building urban resilience open up avenues for effective urban climate governance, and as the case studies from low- and middle-income countries below demonstrate, these can be initiated at a number of different levels and through a variety of approaches.

Emerging lessons for mainstreaming urban CBA

CBA is one of multiple tools for municipal climate governance

The Municipality of eThekwini (the local government responsible for the city of Durban, South Africa) recognizes that CBA is one of the several types of tools at its disposal to address climate change, and is adopting a variety of approaches to adaptation and mitigation. The adaptation initiatives of eThekwini Municipality are described elsewhere in urban climate governance literature (Carmin et al., 2012; Roberts, 2008; Roberts et al., 2012), with the city seen as a global leader in urban adaptation. The city's climate protection programme has been driven by the Environmental Planning and Climate Protection Department, with an emphasis on municipal-, ecosystem-based adaptation, and CBA, and has also focused on developing new tools and mainstreaming climate protection into city plans and operations. Seven years of adaptation action have demonstrated that there is no textbook approach to integrating CBA with other adaptation activities, in the context of multiple pressures including social inequity, substantial infrastructure backlogs, and risk-tolerant communities, which represent a 'wicked' mix of problems to which climate change is added (Roberts, 2013). The municipality clearly recognizes that approaches to mainstreaming CBA are not without difficulties and that a learning-by-doing approach is necessary.

The difficulties of developing a participatory approach to adaptation were highlighted with the creation of the Durban Climate Change Partnership (DCCP) in 2010, with a steering committee representing all major stakeholder groups being established (Roberts, 2013). However, due to a combination of factors, including distrust between parties, a lack of sustained motivation and influential leadership, and the related inability to find funding external to local government, by August 2012, the activities of the DCCP had stalled despite the local government's willingness to encourage a participatory approach. Local government was prevented from continuing to fund the partnership by national financial regulations and the lack of a legal structure which could equally accommodate the various stakeholder groups and eThekwini Municipality.

Another lesson which has emerged is the importance of local champions. At the municipal level, Debra Roberts, head of the Environmental Planning and Climate Protection unit, is recognized as the champion making Durban a leader in adaptation (Carmin et al., 2012). In the Buffelsdraai Community Reforestation Project, where local households grow native tree seedlings in exchange for credit notes, one of the local facilitators championed sponsoring school fees using credit notes, with resultant beneficial impacts on the local community. At the same time, analysis showed that many 'tree-preneurs' were not aware of the motivation behind the reforestation project (Debra Roberts, personal communication, November 28, 2013), and thus, if an eco-system-based adaptation approach is to affect local understanding of climate change impacts, this needs to be at the forefront. However, this also highlights the difficulties of making adaptation a priority in a context of social and economic deprivation where more immediate needs may be prioritized by residents.

The Durban case shows that a single approach to adaptation is insufficient, and CBA needs to be integrated as part of a package of tools applied at the city level. In order to effectively mainstream a CBA agenda, a city needs to ensure that any new institutions established appropriately respond to the needs of stakeholders and are suitably framed and facilitated so as to break down any existing distrust which may be a result of the existing political and social power relations.

Ensure top-down priorities are aligned with local-level needs

While the Durban example illustrates the role of city-level policies, the case of Guwahati in India demonstrates that, in order to avoid mismatched priorities across different levels of government (Leck & Simon, 2012), a coherent top-level policy framework is important in shaping the mainstreaming of CBA into local- and state-level planning. Because urban development in the Indian federal system is a State subject, State Government is a powerful entity in facilitating the urban climate adaptation agenda. While Central Government has the powers to make policies and schemes related to urban development, their adoption and implementation rests with the State. However, the State Climate Change Action Plans, which all States should develop and implement under the Prime Minister's National Action Plan on Climate Change (NAPCC), and the National Mission on Sustainable Habitat, one of the flagship Missions of the NAPCC (yet to be implemented), are the only windows of opportunity for integrating CBA into urban development planning processes. The Indian Constitution's 74th Amendment Act gives powers to urban local bodies to plan for themselves, and requires cities to elect mayors and ward councillors. However, mechanisms for elected representatives to receive regular feedback from the community are ill-defined (Sivaramakrishnan, 2007). Elected representatives are often unfamiliar with approaches to addressing climate change and its

impacts at the city level (Divya Sharma, personal communication, December 30, 2013). Vested interests play a critical role in shaping action, as does the character of the political economy of the State in terms of reigning party and political–bureaucratic relationships. However, if implemented in its true spirit, the 74th Amendment Act holds potential for supporting locally led approaches to climate change through decentralization of functions to urban local bodies.

For communities, their ability to demand action on disaster-risk reduction (DRR) and climate adaptation planning are limited by a lack of awareness. There is no system of community feedback or participation (both core components of CBA) within city development planning and State governments' budgetary planning, preventing formal and regular communication between citizens and the government (TERI, 2011).

A study by The Energy and Resources Institute (TERI, 2011), in the context of the Asian Cities Climate Change Resilience Network (ACCCRN)[1] initiative, developed a detailed framework for the preparation and mainstreaming of a resilience strategy for Guwahati. The resilience strategy consisted of an integrated plan involving housing, ecologically sensitive urban planning, and urban infrastructure and services, overarched by DRR. This was substantiated by a detailed regulatory and institutional analysis of the urban development planning processes adopted in Guwahati and the State of Assam, to identify the entry points for mainstreaming the strategy.

The Guwahati study demonstrated that adaptation actions can be developed if the city government has the will and leadership needed to take up the tasks and actions towards climate resilience, and where knowledge and capacity gaps are filled by external actors like the national ACCCRN partners (see also Kernaghan & da Silva, 2013). However, implementation of strategies not falling under the 'regular' mandate of the city or the state government remains questionable. While the Guwahati city government has accepted the study recommendations, to adopt these they need State Government's acceptance of the resilience strategy to be forwarded to the city for implementation. This is logical given the regulatory-institutional set-up where urban development is a State subject, and both the State and national levels should work with cities when building climate policy which reframes city-level development (Corfee-Morlot, Cochran, Hallegatte, & Teasdale, 2011; Sharma & Tomar, 2010). Adoption of CBA would require institutional mandates and proper application of existing regulations for community participation. Additionally, planning for any new challenge, particularly for short- and medium-term actions, requires the solution to be based on existing implementation frameworks of the state or city. The long-term actions could focus on bringing in the absent laws, regulations, and policies that could steer and sustain such efforts for other cities.

Seize opportunities for institutional reform to put community-based approaches on the agenda

In South America, efforts to reshape policies and regulations are emerging. Case studies from Argentina and Uruguay show how certain local governments are seeking to engage in multilevel adaptation planning, and how these approaches may be strengthened by integrating community-based perspectives in a two-way process. A research project analysing the impact of climate change on coastal areas of the Rio de la Plata, and potential for multilevel governance of risk-management and adaptation, compares two cities each in Uruguay and Argentina (Almansi & Hardoy, 2013). This study offers a number of insights into the socio-institutional barriers which are hampering full integration of community voices in urban climate governance[2].

Uruguay is undergoing an institutional redesign, with a mandate to use more holistic approaches at multiple levels. In 2009, the government created the National Climate Change and Variability Response System for coordinating risk prevention, mitigation, and adaptation, and preparing the national climate change response plan. However, although the new laws consider the cross-cutting nature of climate change and the necessary interaction needed between land use, water use, and the environment and disaster-risk regulations, advances in the national institutional architecture have not yet permeated to sub-national levels, excepting isolated pilot projects. Active citizen participation in drafting of territorial and urban water plans is also being considered, but this requires a strong political will and commitment from technical experts, given the deviation from traditional processes. The pace of change is thus slow.

In Argentina, the process is nascent, without large-scale change in the institutional structures or legal frameworks to support cross-cutting climate change adaptation (Almansi, 2010). However, some advances have been made. The secretariats of different Ministries are working together to integrate DRR and climate change adaptation into territorial planning – however, climate change issues and integrated approaches do not really permeate other Ministries nationally or guide local planning and DRR.

This comparative study highlights the difficulties of DRR and climate change planning, which are seen as the responsibility of the environmental offices, while plans for urban services falling under other offices may actually increase vulnerability and risk if implemented, and may pay no regard to adaptation. Local governments lack financial resources earmarked for disaster risk, much less for climate change planning, nor the associated technical capacity or legal mandate.

On the other hand, communities historically have developed strategies for adaptation to different environmental conditions in anticipation of government actions. Self-organized community groups exist, for example issuing emergency warnings. Climate change now brings new challenges associated with uncertainty and its effects on everyday life. Local community demands may consolidate risk, for example, when informal settlements in flood risk areas are regularized in order to avoid social conflict arising from relocation. In the best cases, this regularization is approved with the requirement of carrying out preventive hydrological works. (Almansi, Hardoy, & Pandiella, 2014).

What emerges from this is an opportunity for raising awareness about the links between urban and infrastructure planning, emergency management, housing, and urban services in the context of increasing climatic uncertainty, and the potential for reshaping socio-institutional relations. There are difficulties in learning from the failures of old land use regulations, plans, and poverty reduction strategies, which may themselves have contributed to the cities' current risks (Almansi et al., 2014). A municipality which does not limit the speculative retention of land, whose building regulations create spatial segregation between rich and poor, and which does not create mechanisms for the recovery and distribution of land rents, is ill-positioned to consider an urban risk reduction policy (Almansi et al., 2014). This presents an opportunity to improve urban planning and poverty alleviation policies from a community base. Local communities can generate clear and practical information and disseminate it, opening different options to enable more effective actions in cities. Involvement by communities which have been affected by climate impacts can drive action by decision-makers, and awareness of other communities to start adaptation, which is particularly important in the context of uncertain or non-existent data, and reluctance by politicians to share information about probable risk areas. As Agrawal et al. (2009) suggest, these communication barriers hamper effective multi-level governance by blocking participation by local organizations in the local government response.

Institutionalise multi-stakeholder approaches to facilitate national mainstreaming

In Indonesia, city-level adaptation approaches are an opportunity to shape governance at different scales, by scaling up multi-stakeholder approaches to urban climate change planning from two cities (Bandar Lampung and Semarang) to another six cities. This is being done through a multi-pronged process, making use of local and national networks of city-level actors through the institutionalization of multi-stakeholder groups, strengthening the take-up of a resilience-building approach within and beyond the city, through direct engagement with national state actors (Syam, 2013). As part of ACCCRN, city resilience strategies for Bandar Lampung and Semarang were developed through an inclusive, multi-stakeholder process building on vulnerability assessments. These are now being mainstreamed into city development plans. The process rests on having a wide cross-section of stakeholders, including civil society and NGOs, academics, and practitioners, working with a range of city-level officials, within a city team formally agreed with the city government, thus giving it the necessary institutional standing. A number of other bodies have been established to institutionalize the process at different scales (Table 1), thus facilitating uptake nationally, and demonstrating how adaptation processes need to be 'multidimensional and multiscalar', bringing together 'actions, actors, sectors, and governance levels' (Leck & Simon, 2012).

In Semarang, a body called the Initiative for Urban Climate Change and Environment (IUCCE) supports the resilience-building process at the city level, bringing together different stakeholders to coordinate local processes and gather evidence. The establishment of the IUCCE forms part of the strategy for sustaining city-level activities post-ACCCRN, providing opportunities for stakeholders from multiple sectors to assist the city government. Meanwhile, the Best Practice Transfer Program supports replication by other Indonesian cities through city-to-city peer-learning opportunities. Nationally, the Indonesian Climate Alliance brings together local and national government, civil society, donors, academics, and private sector representatives to actively support the institutionalization of urban climate resilience. The Indonesian Climate Alliance informed the development of the Indonesian Climate Change Adaptation plan, which is being formally adopted by the National Development Planning Agency. The recommendations include a specific mandate for local governments to develop their own local climate change adaptation plan as a downscaling strategy.

These initiatives, spurred on by ACCCRN, demonstrate a growing momentum for replicating and geographical scaling up through national engagement, embedding urban climate planning more deeply within cities and nationally. Building alliances between national and city actors can incentivize replication and scaling up of urban

climate interventions. Establishing dedicated institutions can synchronize perspectives between the national and city levels. Standardized methodologies and tools can facilitate this, particularly vulnerability assessment and city resilience strategy processes, as these initial steps secure the engagement of a range of actors from communities to officials.

Participatory research to influence local government adaptation planning

Where community-level knowledge may be acknowledged by local government actors, the approaches to adaptation they propose may not be – and participatory research approaches can facilitate local knowledge to shape adaptation actions. This is the case of Quy Nhon, a Vietnamese coastal city in Binh Dinh province, where planned urbanization and infrastructure development has begun encroaching on low-lying agricultural land on the city outskirts (DiGregorio, 2013a; DiGregorio, 2013b). In 2009, typhoon Mirinae killed seven people and caused roughly $21 million USD in damage in Quy Nhon City. Mirinae is the type of extreme hydro-meteorological event projected to worsen in future. DiGregorio and Huynh (2012) examined the causes of Mirinae's severity, in order to reduce risk and prevent such future catastrophes, particularly concerning plans to urbanize the Ha Thanh River delta, the area most severely impacted by the flood.

Through interviews at 21 sites in the delta, the researchers determined historical flood patterns and local adaptation strategies, the height and chronology of flooding, and developed hypotheses regarding causes of the flood's severity. They tested these hypotheses by reviewing official damage assessments, urban plans, satellite imagery, and history of infrastructure construction and urbanization in the delta. A senior researcher[3] at the Southern Institute of

Table 1. Institutionalizing urban climate change resilience within urban governance in Indonesia.

Goal: Institutionalising urban climate change resilience within the Government of Indonesia, including the incorporation of resilience-building strategies in planning and budgeting processes across all sectors and levels of government.

Objectives	Platform(s)	Key actor(s)
Build active national-level multi-stakeholder platform to support institutionalization of urban climate change resilience	Indonesia Climate Alliance	Representatives from government ministries and departments, donors, selected civil society, and private sector
Provide knowledge and toolkits for national government to integrate and apply urban resilience strategies	Online vulnerability assessment tool and system; climate spatial planning	Government ministries and agencies, universities
Work with 'early-adopter' cities and replication cities to apply methodologies and build knowledge to support momentum for national mainstreaming of urban climate resilience	Best practice transfer programme; Initiative for urban climate change and environment (IUCCE) and network (IUCCN)	City-level teams from both early-adopter and replication cities, universities
Advocate for funding mechanisms for climate change adaptation at the national level	Indonesia Climate Alliance, existing climate financing mechanisms	Representatives from government ministries and departments, donors, selected civil society, and private sector

Water Resources Research developed a whole watershed hydrological model, to test for impacts related to recent construction in the delta, full implementation of the proposed Nhon Binh Area Plan, and full implementation plus climate change under conditions of a Mirinae-type storm.

Interviews with residents in the Ha Thanh delta revealed that recent urbanization and infrastructure construction may have aggravated the flood. Floods are a seasonal occurrence in the delta to which generations of inhabitants have adapted. Older residents realized that new construction in the delta changed flood patterns over the years, resulting in higher flooding overall. This was confirmed by hydrological modelling. Before 2003, the relatively few barriers in the Ha Thanh delta allowed floodwater to gradually dissipate. New urban areas and infrastructure constructed between 2003 and 2009 constrained floodwater flows. The study found that if the Area Plan for Nhon Binh ward was fully implemented, damage from seasonal flooding would increase in areas outside the plan, and loss of life and property from an extreme climate event would be much greater across the delta. Under the approved climate scenario to 2050, with flood conditions similar to Mirinae, fully implementing the Area Plan could result in higher flood levels, with the most severe impacts suffered by residents of older settlements.

Following the advice of residents and hydrological modelling, the researchers concluded that improvement of drainage rather than dike construction should be the primary means of reducing vulnerabilities to flooding. This contrasts the civil engineering approach that attempts to balance the risk of flooding against perceived economic benefits of urbanization. Information from this research was shared through workshops with the provincial Department of Construction, as part of ACCCRN activities in the city. Despite this, the Department chose to reject a 'room for rivers' approach to urbanization of the delta in favour of recommendations offered by water engineers working for a national institute, allowing for greater urbanization of the Ha Thanh River floodplain. Based on the research findings, this decision will likely lead to the need for construction of hugely costly flood diversion channels. At the same time, as a result of this study, provincial administrators have begun reassessing plans to incorporate more flood plain areas into the growing city (Centre for Urban Planning and Construction Inspection, 2012). On one hand, they are looking to expand towards the hills, whilst on the other, they have used several key planning elements arising out of the community-based research to consider development of existing settlements or 'urban clusters' in floodplains, with flood channels and agricultural flood buffer zones. As the city revises its master plan to 2030, this urban clusters approach is appearing as a key element in planning for climate adaptation.

The Quy Nhon case demonstrates the potential for local knowledge in informing urban development plans, and the value of research to confirm and quantify community observations. However, it also highlights the tensions between social, physical, and economic impacts of rapid urban development on local communities, and pressures faced by local officials to accommodate developers and the potential short-term economic benefits they offer (Brown, Dayal, & Rumbaitis del Rio, 2012).

Addressing obstacles to mainstreaming urban CBA

The case studies above have provided insights into the different scales at which CBA can be mainstreamed in urban contexts, and the different ways in which this is happening, as well as possibilities for geographical scaling up nationally. Multilevel governance, with national and local governments working together on urban climate governance, has been emphasized in the literature (Bulkeley & Tuts, 2013; Corfee-Morlot et al., 2011), and by extension, the multilevel approach should extend to local-level community and civil society organizations. If urban resilience is to enter the discourse at the city scale, the capacity of city residents as well as officials to understand the implications of climate change and how to adapt to it needs to be strengthened. Where certain actions are beyond the scope of individuals or community collectives, such as putting in storm drains, opportunities for scaled up action by the state should be highlighted and citizens should be empowered to demand action by the government. This can be facilitated by mainstreaming climate change considerations into existing processes of consultation and planning, and cities planning for adaptation will need to engage differently with communities (Bulkeley & Tuts, 2013).

The case studies have demonstrated that while CBA is not the only approach to addressing climate change in urban areas, it is a valuable tool within a package of tools to be applied by cities. CBA is very relevant given that climate change impacts are highly context specific and thus should be informed by local knowledge and experience. However, in order to be applied effectively and supported on a city-wide scale, CBA should be mainstreamed alongside other climate interventions, while recognizing that different areas will have their specific adaptation actions. Thus, the intervention of external actors, such as NGOs, to facilitate the link between the communities and local government, may be necessary, particularly where communities are not easily identifiable or defined. This was demonstrated in Quy Nhon where a research project gathered the voices of grassroots groups and validated their local observations through scientific modelling before linking them to local government planning processes. Similar roles could be played by civil society organizations. It is important to note that an external donor-driven process may provide the initial impetus for such approaches, such as the ACCCRN process encouraging multi-stakeholder engagement in vulnerability assessments and development of city resilience strategies, while in Uruguay, a research project facilitated local community awareness of and engagement in climate change planning. On the other hand, mainstreaming may be initiated from the top-down, such as in Durban where community-based approaches are actively applied alongside other adaptation mechanisms, with built-in poverty alleviation and education strategies, thus mainstreaming adaptation across sectors.

Mainstreaming of CBA should be facilitated if there is an existing policy framework which allows climate change considerations to be mainstreamed across the board. Policy frameworks which require community consultation or participatory approaches can facilitate the inclusion of local community groups in climate change planning processes; however, even where such frameworks exist, as in India, communication barriers, a lack of locally available and understandable data, and local power relations can hamper effective participation. Similarly, there are challenges to ensuring participation happens through a process of deliberative governance, rather than as part of a preordained agenda (Ayers, 2011). New institutional forums operating on urban scales may be required (Bulkeley & Tuts, 2013) and could strengthen accountability and inclusiveness by devolving authority to the local level. Where local communities are well-informed and able to effectively participate in and shape local planning processes, they can hold local bodies to account, and this can be the beginning of a transformative process of social and political change.

The examples presented above demonstrate some of the benefits emerging out of specific attempts to mainstream community-based approaches in urban contexts. In fast growing urban areas, where local communities may face a number of development challenges, CBA can be integrated alongside a wider development package with positive effects on education and livelihoods. Community-based approaches can have an empowering effect, particularly where mechanisms are developed for community voices to feed into planning processes as in Quy Nhon, where development pressures might otherwise outweigh grassroots' rights, or in Indonesia where community vulnerability assessments shaped city resilience strategies. Where opportunities arise, they should be seized to facilitate community engagement with local officials, as in the case of Uruguay's national policy change opening up space for local reshaping of institutional frameworks to build forms of resilience that support transformation towards inclusive governance, rather than the rigidity of the status quo (Pelling & Manuel-Navarrete, 2011).

Nevertheless, obstacles remain in attempts to mainstream urban CBA in city- and national-level planning, and in enabling truly multilevel governance. Capacity gaps remain at multiple levels: in understanding the potential impacts of climate change and the best response, within both local government and local communities; in accessing urban adaptation financing; and in the ability of intermediary organizations to support local communities. Existing institutional structures may shape or prevent inclusive approaches or may be constrained by the lack of appropriate legal mandates to enable action, as with the DCCP in Durban. In other contexts, such as the Latin American cases, the climate change agenda may be hampered by the persistence of an emergency response and good development discourse, rather than forward-looking transformative planning which could encompass all of these approaches.

More fundamentally, in diverse and dynamic urban contexts, it may be difficult to clearly define 'communities' or tensions and power imbalances may exist within communities (Forsyth, 2013). Participation may be driven by reasons other than an understanding of the need for adaptation, as with the 'tree-preneurs' in Durban. There may also exist tensions between local communities and economic development needs and demands, as in Quy Nhon, whether driven by the government, private developers, or both. Therefore, adaptation may be seen not as something to be done in order to become a resilient city, but rather something necessary in order to achieve other city development goals (DiGregorio, 2013a) – and the city's development vision may affect the manner in which local communities are included, or not, in this process.

'Adaptation for whom' should remain a central question to be addressed in planning processes. The matter of who defines the climate-related problem becomes crucial in determining the type of solutions proposed. The case studies demonstrate that climate change adaptation is very much a 'wicked' problem of governance, and as a consequence requires that power relations and equity issues be addressed, beyond solely applying technocratic or cost-versus-benefits approaches (Dewulf, 2013). There are lessons to be learnt from community-led approaches to development, such as the work of community federations in African and Asian cities (see for example, Mitlin & Satterthwaite, 2012), particularly given the difficulty of separating adaptation from development (Cannon & Muller-Mahn, 2010).

The five examples offer some lessons for overcoming obstacles. The overarching policy framework, at national or state level, plays a key role in shaping action at the local level through legal mandates, as in India. However, the institutional framework itself is not enough if there is insufficient capacity in local bodies to ensure policies and approaches are implemented in the appropriate manner or if local power relations distort incentives. Better communication between local communities and their representatives can facilitate a discourse around resilience. Organized communities may be better placed to push forward an agenda, and may be organized around a common hazard, or supported by external intervention by NGOs or researchers. This is aided by improved awareness at the community level of climate change risks and actions which can be taken to address these. There is a role here for strategically located champions to drive the adaptation agenda, whether at the community or government level or outside these sectors. Thus, while mainstreaming, CBA can facilitate the application of community-based approaches on a wider geographical scale, localized drivers on the ground are still required for its implementation. Nevertheless, it remains the case that there is no 'one-size-fits-all' solution, and in planning activities, it is important to be aware of local priorities, which may not be climate change, and thus integrate adaptation into the pre-existing development agenda.

Conclusions

Climate change adaptation provides an opportunity for transformative change in socio-institutional structures at the national and local scale (Pelling, 2011), and mainstreaming community-based approaches is one avenue for this change by giving agency to the local level. Adaptation actions as demonstrated by local communities can complement actions by local government, within a larger toolbox of responses. However, mechanisms are required to facilitate local community voices being heard in planning processes, whether through institutional reforms supporting participatory planning and recognizing this at the national as well as local scales or research feeding into planning processes, and to empower local communities to 'take charge of the direction of change' (Norris, Stevens, Pfefferbaum, Wyche, & Pfefferbaum, 2008, p. 143). Recognizing that for community-based approaches to lead to transformative change, they should be combined with other mechanisms to support the ability of local communities to contest existing power relations (Dodman & Mitlin, 2013) and hence engage in the 'adaptive challenge' that questions the creation of current systems and structures (O'Brien, 2012).

What emerges from the case studies is a clear lesson that communities can provide valuable insight into city development plans and activities addressing adaptation beyond the community scale, and there is a role for building the capacity of local organizations and government bodies to ensure that these views are captured and taken into consideration when planning for adaptation. While climate change adaptation approaches have to be context specific, this paper has used lessons from a growing number of urban experiences to demonstrate five emerging opportunities behind mainstreaming CBA. At the same time, CBA can bring new adaptation methods and local knowledge to improve city- and national-level policies and approaches. CBA is also a mechanism for ensuring that the stresses and risks associated with climate change are considered in an integrated manner alongside other problems currently faced by local populations as part of a development-based approach to adaptation.

Acknowledgments

The authors would like to thank the anonymous reviewers for their constructive comments, as well as Dr Hannah Reid for her comments. This paper arose out of the CBA7 Conference 2013 on Mainstreaming CBA, and draws on the presentations made in the plenary session on CBA in Urban Areas, which was sponsored by the Rockefeller Foundation.

Notes

1. The Asian Cities Climate Change Resilience Network (ACCCRN) is an eight-year initiative funded by the Rockefeller Foundation, which aims to demonstrate approaches to building urban resilience to climate change in secondary Asian cities. Initiated in 10 cities across Thailand, Vietnam, Indonesia, and India, the programme is now extending to Bangladesh and the Philippines.
2. See also the project website http://www.iied-al.org.ar/riberas/home.html
3. To Quang Toan.

References

Agrawal, A., Kononen, M., & Perrin, N. (2009). *The role of local institutions in adaptation to climate change* (The Social Development Papers paper number 118). Washington, DC: The World Bank.

Almansi, F. (2010). The relationship between disaster risks and urban planning in Argentina, Case study prepared for ISDR Global Assessment Report 2011. In C. Johnson. *Creating an enabling environment for reducing disaster risk: Recent experience of regulatory frameworks for land, planning and building in low and middle-income countries.* Global Assessment Report in Disaster Risk Reduction, ISDR. Retrieved January 2, 2014, from http://www.preventionweb.net/english/hyogo/gar/2013/en/bgdocs/Johnson,%202011.pdf

Almansi, F., & Hardoy, J. (2013). *Climate change impact in coastal areas of the Rio de la Plata River: Actions for prevention and adaptation.* Presentation at CBA 7 conference, Dhaka, April 2013.

Almansi, F., Hardoy, J., & Pandiella, G. (2014). Impacto del cambio climático en ciudades costeras del estuario del Rio de la Plata. *Medio Ambiente y Urbanización, 79,* 217–237.

Anguelovski, I., & Carmin, J. (2011). Something borrowed, everything new – current opinion: Innovation and institutionalisation in urban climate governance. *Environmental Sustainability, 3*(3), 169–175.

Ayers, J. (2011). Resolving the adaptation paradox: Exploring the potential for deliberative adaption policy-making in Bangladesh. *Global Environmental Politics, 11*(1), 62–88.

Ayers, J., & Forsyth, T. (2009). Community-based adaptation to climate change, strengthening resilience through development. *Environment, 51*(4), 23–31.

Breil, M., & Johnson, K. (2012). Conceptualizing urban adaptation to climate change. *Review of Environment, Energy and Economics (Re3).* Doi: 10.7711/feemre3.2012.08.001.

Brown, A., Dayal, A., & Rumbaitis del Rio, C. (2012). From practice to theory: Emerging lessons from Asia for building urban climate change resilience. *Environment and Urbanization, 24*(2), 531–556.

Bulkeley, H., & Tuts, R. (2013). Understanding urban vulnerability, adaptation and resilience in the context of climate change. *Local Environment, 18*(6), 646–662.

Cannon, T., & Mueller-Mahn, D. (2010). Vulnerability, resilience and development discourses in context of climate change. *Natural Hazards, 55*(3), 621–635.

Carmin, J., Anguelovski, I., & Roberts, D. (2012). Urban climate adaptation in the global South: Planning in an emerging policy domain. *Journal of Planning Education and Research, 32*(1), 18. Doi:10.1177/0739456X11430951.

Centre for Urban Planning and Construction Inspection (2012). *Đề cương lập nhiệm vụ điều chỉnh tổng thể quy hoạch chung xây dựng thành phố Quy Nhơn đến năm 2030, định hướng đến năm 2050.* [Terms of reference for the revision of the master construction plan for Quy Nhon city to 2030, with a view to 2050]. Quy Nhon, Bình Định: Department of Construction.

Corfee-Morlot, J., Cochran, I., Hallegatte, S., & Teasdale, P. (2011). Multilevel risk governance and urban adaptation policy. *Climatic Change, 104*(1), 169–197.

Dewulf, A. (2013). Contrasting frames in policy debates on climate change adaptation. *WIREs Climate Change, 4*(4), 321–330. Doi:10.1002/wcc.227.

DiGregorio, M. (2013a). *Planning for climate resilience: Learning from grassroots*, Presentation at CBA 7 conference, Dhaka, April 2013.

DiGregorio, M. (2013b). Learning from Typhoon Mirinae: Urbanization and climate change in Quy Nhon City, Vietnam. ISET. Retrieved May 25, 2013, from http://www.acccrn.org/sites/default/files/documents/ISET_LearningFromTyphoonMirinae_Final_130419.pdf

DiGregorio, M., & Huynh, C. V. (2012). Living with floods: A grassroots analysis of the causes and impacts of Typhoon Mirinae. Retrieved May 25, 2013, from http://www.acccrn.org/resources/documents-and-tools/2012/07/01/living-floods-grassroots-analysis-causes-and-impacts-typhoo

Dodman, D., & Mitlin, D. (2013). Challenges for community-based adaptation: Discovering the potential for transformation. *Journal of International Development, 25*(5), 640–659. Doi:10.1002/jid.1772

Dodman, D., & Satterthwaite, D. (2008). Institutional capacity, climate change adaptation and the urban poor. *IDS Bulletin, 39*(4), 67.

Few, R., Brown, K., & Tompkins, E. L. (2007). Public participation and climate change adaptation: Avoiding the illusion of inclusion. *Climate Policy, 7*(1), 45–59.

Forsyth, T. (2013). Community-based adaptation: A review of past and future challenges. *WIREs Climate Change.* Doi:10.1002/wcc.231

Hurlimann, A. C., & March, A. P. (2012). The role of spatial planning in adapting to climate change. *WIREs Climate Change, 3*(5), 477–488.

IPCC (2007). Climate change 2007: Appendix to synthesis report. In A. P. M. Baede, P. van der Linden, & A. Verbruggen (Eds.), *Climate change 2007: Synthesis report. Contribution of working groups I, II and III to the fourth assessment report of the intergovernmental panel on climate change* (pp. 76–89). Geneva: IPCC.

Jabeen, H., Johnson, C., & Allen, A. (2010). Built-in resilience: Learning from grassroots coping strategies for climate variability. *Environment and Urbanization, 22*(2), 415–431.

Jahan, R. (1995). *The elusive agenda: Mainstreaming women in development.* London: ZED Books.

Kernaghan, S., & da Silva, J. (2013). Initiating and sustaining action: Experiences building resilience to climate change in Asian cities. *Urban Climate* http://dx.doi.org/10.1016/j.uclim.2013.10.008

Leck, H., & Simon, D. (2012). Fostering multiscalar collaboration and co-operation for effective governance of climate change adaptation. *Urban Studies.* Doi:10.1177/0042098012461675

Leichenko, R. (2011). Climate change and urban resilience. *Current Opinion in Environmental Sustainability, 3*(3), 164–168.

Mitlin, D., & Satterthwaite, D. (2012). Addressing poverty and inequality – new forms of urban governance in Asia. *Environment and Urbanization, 24*(2), 395–401.

Norris, F., Stevens, S., Pfefferbaum, B., Wyche, K., & Pfefferbaum, R. (2008). Community resilience as metaphor, theory, set of capacities, and strategy for disaster readiness. *American Journal of Community Psychology, 41*(1–2), 127–15.

O'Brien, K. (2012). Global environmental change II: From adaptation to deliberate transformation. *Progress in Human Geography, 36*(5), 667–676.

Olhoff, A. & Schaer, C. (2010). Screening tools and guidelines to support the mainstreaming of climate change adaptation into development assistance – a stocktaking report. New York: UNDP.

Pelling, M. (2011). *Adaptation to climate change: From resilience to transformation.* London, UK: Routledge.

Pelling, M., & Manuel-Navarrete, D. (2011). From resilience to transformation: The adaptive cycle in two Mexican urban centers. *Ecology and Society, 16*(2), 11.

Pervin, M., Sultana, S., Phirum, A., Camara, I., Nzau, M., Phonnasane, V., … Anderson, S. (2013). *A framework for mainstreaming climate resilience into development planning.* IIED Working Paper, IIED, London

Reid, H., Alam, M., Berger, R., Cannon, T., Huq, S., & Milligan, A. (2009). Community-based adaptation to climate change: An overview. *Participatory Learning and Action: Community Based Adaptation to Climate Change, 60*, 11–33.

Roberts, D. (2008). Thinking globally, acting locally – institutionalising climate change at the the local government level in Durban, South Africa. *Environment and Urbanisation, 20*(2), 521–537.

Roberts, D. (2013). *Busting the Myths – An African reality check on Community Based Adaptation in Durban, South Africa.* Presentation at CBA 7 conference, Dhaka, April 2013.

Roberts, D., Boon, R., Diederichs, N., Douwes, E., Govender, N., McInnes, A., & McLean, C. (2012). Exploring ecosystem-based adaptation in Durban, South Africa: "Learning-by-doing" at the local government coal face. *Environment and Urbanization, 24*(1), 167–195.

Roy, M., Hulme, D., & Jahan, F. (2013). Contrasting adaptation responses by squatters and low-income tenants in Khulna, Bangladesh. *Environment and Urbanization, 25*(1), 157–176.

Satterthwaite, D., Huq, S., Reid, H., Pelling, M., & Romero Lankao, P. (2007). *Adapting to climate change in urban areas: the possibilities and constraints in low- and middle-income nations.* Human Settlements Working Paper, IIED, London.

Satterthwaite, D., & Mitlin, D. (2014). *Reducing urban poverty in the global south.* Abingdon: Routledge.

Sharma, D., & Tomar, S. (2010). Mainstreaming climate change adaptation in Indian cities. *Environment and Urbanization, 22*(2), 451–465.

Sivaramakrishnan, K. C. (2007). Democracy in urban India. Urban Age. Retrieved January 15, 2014, from http://lsecities.net/media/objects/articles/democracy-in-urban-india/en-gb/

Soltesova, K., Brown, A., Dayal, A., & Dodman, D. (2014). Community participation in urban adaptation to climate change: Potential and limits for CBA approaches. In L. Schipper, J. Ayers, H. Reid, S. Huq & A. Rahman (Eds.),

Community-based adaptation to climate change: scaling it up (pp. 214–225). Abingdon, UK: Routledge.

Syam, D. (2013). *National engagement and replication in Indonesia.* Presentation at the CBA7 conference, Dhaka, April 2013.

TERI (2011). *Mainstreaming urban resilience planning in Indian cities: A policy perspective.* Report prepared for ACCCRN in India. Retrieved May 25, 2013, from http://www.acccrn.org/ sites/default/files/documents/Final_Mainstreaming%20Urban %20Resilience%20Planning%20copy.pdf

Twigg, J. (2007). Characteristics of a disaster-resilient community, a guidance note. Retrieved May 25, 2013, from http:// practicalaction.org/docs/ia1/community-characteristics-en-lowres.pdf

Tyler, S., & Moench, M. (2012). A framework for urban climate resilience. *Climate and Development, 4*(4), 311–326.

REVIEW ARTICLE

A review of decision-support models for adaptation to climate change in the context of development

John Jacob Nay[a], Mark Abkowitz[b], Eric Chu[c], Daniel Gallagher[d] and Helena Wright[e]

[a]Program in Integrated Computational Decision Science and Institute for Energy and Environment, Vanderbilt University, USA; [b]Department of Civil and Environmental Engineering, Vanderbilt University, USA; [c]Department of Urban Studies and Planning, Massachusetts Institute of Technology, USA; [d]Adaptation Fund Board secretariat, USA; [e]Centre for Environmental Policy, Imperial College, London, UK

In order to increase adaptive capacity and empower people to cope with their changing environment, it is imperative to develop decision-support tools that help people understand and respond to challenges and opportunities. Some such tools have emerged in response to social and economic shifts in light of anticipated climatic change. Climate change will play out at the local level, and adaptive behaviours will be influenced by local resources and knowledge. Community-based insights are essential building blocks for effective planning. However, in order to mainstream and scale up adaptation, it is useful to have mechanisms for evaluating the benefits and costs of candidate adaptation strategies. This article reviews relevant literature and presents an argument in favour of using various modelling tools directed at these considerations. The authors also provide evidence for the balancing of qualitative and quantitative elements in assessments of programme proposals considered for financing through mechanisms that have the potential to scale up effective adaptation, such as the Adaptation Fund under the Kyoto Protocol. The article concludes that it is important that researchers and practitioners maintain flexibility in their analyses, so that they are themselves adaptable, to allow communities to best manage the emerging challenges of climate change and the long-standing challenges of development.

1. Introduction

For a number of reasons, climate change poses additional negative implications for developing countries as well as poverty-affected communities residing anywhere (Stern, 2006). Poverty is associated with less economic, political, and organizational capacity to adapt, which makes individuals and communities more vulnerable to economic and climate shocks (Dodman & Satterthwaite, 2008). Moreover, poverty-affected communities may live in more vulnerable areas because these are historically the more marginalized areas, which are often the by-products of informal land tenure, a lack of public services, and exposure to natural hazards. Developing economies depend more on climate-sensitive activities, such as rain-fed agriculture, that are more impacted by climate variability (Hertel & Rosch, 2010). Financially constrained governments are less able to devote significant amounts of capital to "climate-proof" infrastructure and improve weather forecasting. Equity concerns such as these must serve as a backdrop for climate adaptation policy (Parks & Roberts, 2010; Shepard & Corbin-Mark, 2009).

Proposed adaptation interventions may generate benefits independent of climate change concerns (Carter et al., 2007). The Fourth Assessment Report of the Intergovernmental Panel on Climate Change (IPCC, 2007) concluded that planned adaptations to climate risks are "most likely to be implemented when they are developed as components of (or as modifications to) existing resource management programs or as part of national or regional strategies for sustainable development." Many general development activities, such as creating more effective and equitable agricultural markets or diversifying livelihood options beyond rain-fed cultivation, can simultaneously improve the lives of the poor and reduce climatic risks.

Climate adaptation mainstreamed into development planning can address pressing global issues such as inequality and natural resource mismanagement through streamlining and supporting existing decision-making processes across different sectors (Halsnæs & Traerup, 2009; Huq & Reid, 2004). "Community-based adaptation" can be interpreted as a field of research and a community of practice rooted in the notion that improving livelihoods and reducing poverty are primary aims, and that adapting to climate change is a means to those ends. Climate change adaptation is "mainstreamed" into development planning to the extent that development plans are predicted to be robust to current climate variability and expected climate change stressors, such as more variable and extreme droughts and floods (Carmin, Dodman, & Chu, 2013).

Recognition of the links between climate change and development has led to the emergence of tools to integrate climate change adaptation into development planning (OECD, 2009; Olhoff & Schaer, 2010; UNDP/UNEP, 2011). The IPCC has called upon researchers to provide "effective approaches for identifying and evaluating both existing and prospective adaptation measures and strategies" (Carter et al., 2007). The need to examine policies has also been highlighted (OECD, 2009) in light of the close links between adaptation and development. Decision-support tools are important for prioritizing adaptation activities that should be scaled up. However, some tools offer limited guidance on the integration of adaptation into planning (Olhoff & Schaer, 2010) and on how local adaptation needs can be matched by international funders. This article intends to support the effort to identify useful tools by reviewing modelling methods, the importance of community engagement and the assessment of costs and benefits, and to shed light on how international financial mechanisms, such as the Adaptation Fund, can benefit from employing such decision-support tools to inform their own funding portfolios. The ultimate goal is to more effectively determine which (if any) development interventions are most likely to improve communities' welfare in light of the expected climatic change.

2. Balancing community input and technical tools

A conceptual issue at the core of this article is a tension between technical tools and community engagement. There is no "one-size fits all" tool or public policy solution. As Ostrom (2007) argued, there are no panaceas for predicting or governing social–ecological systems.

To effectively allocate public expenditures, estimates of costs and benefits, or at least cost-effectiveness, are useful. At the same time, there are examples of capital-intensive projects that may have been technically justified through economic analyses that, in retrospect, have done more harm than good because decision-makers did not understand local realities (Gilligan, Ackerly, & Goodbred, 2013; Haque, 2013). In these cases, if the respective communities had been engaged during the design and review phases, these projects may have been designed differently or abandoned altogether. For example, in the development

of National Environmental Action Plans in Cote D'Ivoire, top-down processes resulted in misidentification of problems and could have wasted limited resources (Ayers, 2011; Bassett & Zuéli, 2000).

Social vulnerability to climate change is a socially constructed phenomenon affected by inequitable resource availability and the entitlements of individuals or groups to call on these resources, including institutional and economic dynamics (Adger & Kelly, 1999). For instance, women's limited access to resources, restricted rights, limited mobility and voice in community, and household decision-making can make them particularly vulnerable to the effects of climate change (Wright & Chandani, 2014). Most technical accounts of "problems" and "solutions" do not take the socially constructed nature of vulnerability into account (Sultana, 2013). Cost-effectiveness measures are not designed to account for non-quantifiable benefits or the issue of *who* benefits. Therefore, researchers, planners, and policy-makers should consider how to strike a balance between fully addressing stated needs of the beneficiaries of an intervention *and* seeking to maximize the intervention's technically derived net benefits or cost-effectiveness.

To assist in assessing future benefits, researchers should develop and utilize tools that facilitate forecasting social, economic and environmental change, and anticipating challenges that may be amenable to intervention from government or civil society. In this article, we call tools of this environmental-economic nature "integrative models." The successful application of integrative models to development planning is to a large extent dependent on the extent to which they are bottom-up. Community input can increase legitimacy of planning and allow the planner to make better predictions. This may ultimately result in more effective interventions.

Before discussing the respective characteristics and relative merits of various types of models and tools, we review two themes important to any approach to mainstreaming adaptation: (1) community participation and engagement and (2) approaching the community or region of interest as a coupled human-natural-engineered complex system.

3. Participation and engagement

Climate adaptation strategies must be implemented at the local level. As a result, community-identified activities are integral to planning. Facilitating public participation and stakeholder engagement is critical to defining climate impacts, understanding local implications, and prioritizing responses. Stakeholder engagement in the design, implementation, and monitoring of interventions is important because the potential impacts of climate change and the actions to reduce these impacts are ultimately interwoven with specific populations and regional vulnerabilities

(Ebi, 2009). Similarly, an area's cultural and local institutional contexts strongly determine the kinds of adaptive strategies people utilize (Adger, Barnett, Brown, Marshall, & O'Brien, 2012; Crate, 2011).

Public participation and engagement processes can strengthen the knowledge and awareness necessary to achieve a sense of citizenship. The idea of citizenship influences the practice and efficacy of participation, the transfer of skills across issues and arenas, and the thickening of alliances and networks (Gaventa & Barrett, 2012). This can also contribute to a broader sense of inclusion of previously marginalized groups within society and potentially increase social cohesion (Gaventa & Barrett, 2012). In this sense, different classes, genders, and cultures play an important role in stakeholder engagement processes (Smith, Vogel, & Cromwell, 2009) and, therefore, also in selecting adaptation strategies (Nielsen & Reenberg, 2010).

If citizen discourse and deliberation play central roles in helping to define impacts and prioritize responses, it must be acknowledged that public discourse and participation in a decentralized political sphere are messy, driven by dynamic, and often contentious, streams of local knowledge (Cheema, 2007), which can all be striving to simultaneously influence institutional change (Hobson & Niemeyer, 2011). Despite this, community-generated knowledge, because of the deliberative processes involved in its creation, can ultimately increase legitimacy of decisions and the likelihood of achieving locally appropriate outcomes (Pringle & Conway, 2012).

Community participation and stakeholder engagement are also keys to facilitating the integration of adaptation and development planning (Halsnæs & Traerup, 2009; Huq & Reid, 2004). The rationale is that adaptation, when addressed simultaneously with other local socioeconomic priorities, can contribute to the livelihoods of people and make improvements in their capacity to deal with climatic change (Halsnæs & Traerup, 2009; Saito, 2012). Local ownership over processes of mainstreaming adaptation into local development can facilitate these programmes' effectiveness (Shaw, 2006), increase their chances for more equitable and just outcomes (Ebi, 2009), and provide opportunities for local innovation (Rodima-Taylor, 2012).

4. Integrative systems approach

In order to understand climatic and non-climatic changes and inform adaptation and development strategies, one must understand the relevant social-ecological systems, as well as their potential critical feedbacks and nonlinear changes (Ostrom, 2009). Human communities and biophysical environments are complex systems with processes operating at nested spatial scales – social units have boundaries such as individual, household, community, and region, whereas biophysical units have boundaries such as patches, stands, forests, watersheds, and biomes (Holling, 2001). Components of both human and biophysical systems are subject to cross-scale interactions and abrupt change (Gunderson, 2010). Communities are characterized by co-evolving social, engineered, and natural systems dynamically affecting one another (Gilligan, Ackerly, & Goodbred, 2013).

Social, natural, and engineered systems co-evolve by interacting in specific places (Gunderson, 2010), for example, infrastructure siting and land-use decisions result in modified physical landscapes. Understanding the dynamics that give rise to effective adaptation requires recognizing how people interact with their environment. A bottom-up analysis of this nature is data hungry (and computationally intensive if modelling and simulation methods are used) and has the difficult task of moving from micro-level details to macro-level patterns and policy recommendations. More top-down methods offer less understanding of the dynamics of coupled systems and are less able to identify how co-evolving factors can determine outcomes (Gilligan, Ackerly, & Goodbred, 2013; Ostrom, 2009). Considering adaptation problems from a social–ecological systems standpoint offers a powerful perspective on the complexity of adaptation, but there are trade-offs in the use of various modelling approaches that explore and simplify this complexity. Various modelling options are, therefore, explored in the following section to present insights into the trade-offs facing decision-makers and researchers.

5. Modelling tools

In the discussion below, we outline a distinction between conceptual and formal models before dividing formal models into those that are equation-based, agent-based, geographic-based, and participation-based. These categories should not be interpreted as rigid or mutually exclusive. Rather, the distinctions are intended to serve as a guide to thinking about general (and compatible) *approaches* to modelling in the adaptation context. We describe these model types abstractly herein; additional detail on each model type and how they are applied in case studies can be found in cited literature.

Models can be used to forecast, illuminate uncertainties, demonstrate trade-offs, and inform policy and planning (Epstein, 2008). Assumptions about important variables of a system and their relationships should be established in any model formulation. This process requires researchers and analysts to test the consistency of (often) previously implicit models and allows the resulting model to be replicated, which facilitates a process of incremental scientific and social learning (Epstein, 2008). We first divide models into those that are more conceptual in nature and those that have a more formalized structure.

5.1. *Conceptual model*

Figure 1 illustrates a broad *conceptual model* for thinking about coupled systems.

This shows the combined social (e.g. economic, regulatory, and informal norms), environmental (e.g. flooding severity and variability), and engineering (e.g. water infrastructure) factors that might lead to unsustainable communities. The conceptual model in Figure 1 focuses on understanding coupled system dynamics by incorporating social and engineering factors. The model is interested in capturing how people adapt to environmental changes under particular institutional and biophysical regimes. It is designed to focus on the complex community or regional system with extant infrastructure and social coping and response mechanisms, and then to investigate possible adaptations.

5.2. *Formal model*

Formal models, which are necessary for more rigorous analyses of conceptual models, can be divided into equation-, agent-, geographic-, and participation-based categories.

5.2.1. *Equation-based*

Both equation-based models (EBM) and agent-based models (ABM) can be deterministic or stochastic, simulate feedback effects, and make extensive use of equations (Bonabeau, 2002). The fundamental difference is that an EBM starts with a set of equations that describes relationships among variables of a system, whereas an ABM starts with behaviours of constituent agents of a system (Parunak, Savit, & Riolo, 1998; Patt & Siebenhuner, 2005). ABMs often have higher computational requirements, which may increase the effort required for sensitivity analyses and calibration. However, EBM approaches may not as sufficiently account for the dynamic processes that can produce macro-level

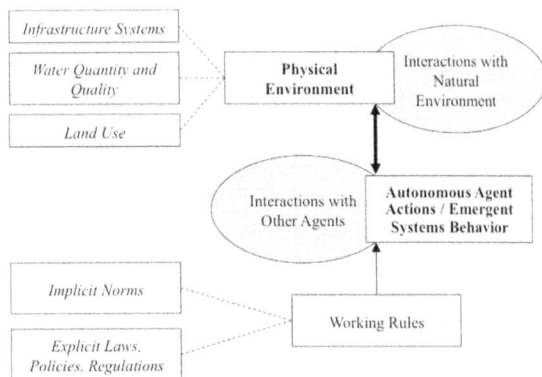

Figure 1.   Conceptual model of the integrated system.

phenomena such as the effect of social norms on individual decision-making, social networks, and heterogeneity in agents' information or control in strategic interactions (Parunak et al., 1998).

EBM adopts a more top-down modelling approach, whereas ABM operates from a bottom-up perspective (Bonabeau, 2002). The top-down approach is often more amenable to making precise predictions and has less parameters to estimate. The bottom-up approach affords flexibility to relax the strong behavioural assumptions of neoclassical economic theory and introduce bounded rationality, social influence, and heterogeneity within a population of simulated economic agents (Filatova, 2009). In many conditions of less than competitive markets where an effective price system is lacking, these additions to micro-behavioural characterizations may allow greater understanding of macro-phenomena than traditional economic models (Poteete, Janssen, & Ostrom, 2010).

Figure 2 illustrates an integrated economic EBM of adaptation to environmental change (based on Fisher-Vanden, Wing, Lanzi, & Popp (2013) integrated assessment model of climate adaptation). The red factors represent exogenous change and the blue factors represent endogenous change. This demonstrates a general practice applicable to all types of formal models: specifying which factors are exogenous and endogenous. *Protective adaptation* (similar to "planned adaptation") shields sectors from impacts, that is, reduces sectors' exposure by reducing the marginal effects of environmental impacts on productivity. Examples of protective adaptation are flood mitigation infrastructure. *Adaptive coping* (similar to "autonomous adaptation") lessens losses that arise once impacts actually affect the sectors in question, that is, increases resilience by lowering the marginal effects of productivity shocks on economic losses. Examples of adaptive coping include migration and changing crop technology. *General equilibrium effects* include relative price changes and substitution responses (Sterner & Persson, 2008). Transformative adaptation, a complete revamp of a social–ecological system in order to become adaptive, would be incorporated in this model as either adaptive coping or protective adaptation. The model moves from conceptual to formal when equations are specified that govern the relationships of the variables connected by arrows.

5.2.2. *Agent-based*

"Agents" in computational ABMs are autonomous decision algorithms that interact with other agents and their environment. Agents can have heterogeneous procedures such as decision-making or learning processes and heterogeneous static (e.g. gender) or dynamic (e.g. wealth and social network) attributes. Social and psychological constraints

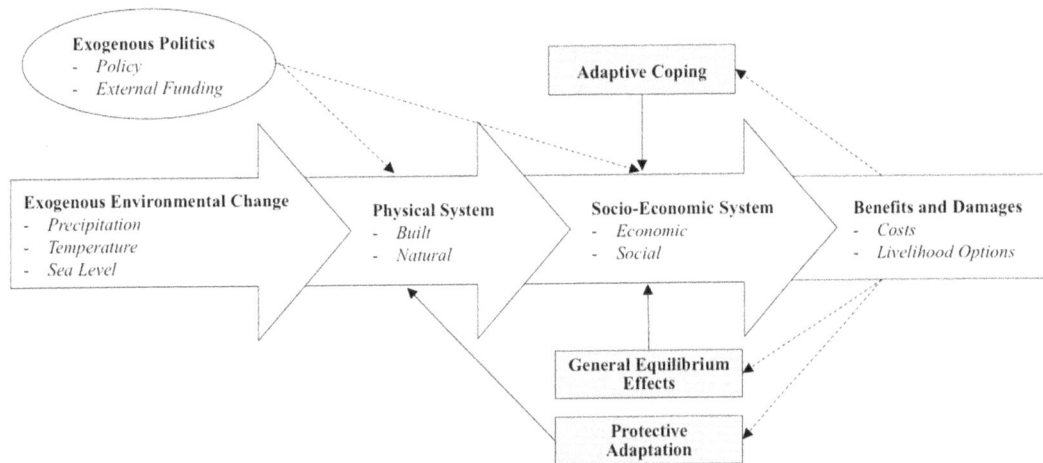

Figure 2. Integrated EBM of adaptation to exogenous change.

can be modelled to affect behaviour. Agent behaviours are actions executed during simulation to achieve some objective(s).

ABMs can be used as tools to explore a range of "what–if" scenarios (Carley, 2002; Lempert, 2002); evaluate how well competing models of human behaviour account for empirical observation (Robinson et al., 2007; An, 2012); better understand potential causal dynamics behind case studies or sequences of events (Janssen & Ostrom, 2006); simulate individuals adapting to changing environments (Balbi & Giupponi, 2009; Patt & Siebenhuner, 2005); and simulate economic, social, and biophysical factors in one integrative model (Schreinemachers & Berger, 2011).

In Patt and Siebenhuner's (2005) review of ABM applied to climate adaptation problems, they note that adaptive capacity is an emergent social phenomenon generated by interaction effects between individuals. They argue that ABM is a suitable tool for understanding adaptive capacity because:

> adaptive capacity arises from a complex system, in which many actions are taken in response to the actions of others. Second, adaptive capacity presents us with a puzzle – maladaptation – that conventional modeling seems unable to solve. Third, it ought to be possible to gather … data necessary to construct valid agent based models of adaptive capacity. Fourth, given the lack of a feasible alternative, ABM may be the only way to predict the success of policy interventions. (p 317)

Community-based adaptation research is focused on agents' adjustments in the face of uncertain change, diffusion of adaptive technologies or behaviours, and decentralized coordination and collective action issues – all phenomena where ABMs have been fruitfully applied (Berger, 2001; Patt & Siebenhuner, 2005). ABMs can represent dynamics in unpredictable systems

(Lempert, 2002) and, uniquely, they can formally accommodate actors' heterogeneity (Abdou, Hamill, & Gilbert, 2001).

No model can consistently predict the behaviour of complex adaptive systems such as coupled social–environmental-engineered community systems (Bradbury, 2002), but ABMs can offer insights into a range of future responses to change and the elements that are most sensitive to those changes. ABM is useful under three conditions (outlined in Patt & Siebenhuner, 2005). First, agent interactions are important for system outcomes, for example, nonlinear changes in system outcomes may result from small changes in agent behaviour. Second, a simpler and more clearly predictive EBM is inadequate (if it is a yet to be experienced phenomenon, such as extreme climate change scenarios, then it might be too difficult to specify the equations that an EBM requires). Third, there are data about agents (EBM does not require micro-level data about agents). The first two conditions hold for most problems of climate adaptation and the third depends on the amount of community participation, relevant theory, and empirical research.

To develop an ABM, one should first select simple attributes of agents and their environments based on existing social science or decision science theory (Cioffi-Revilla, 2010). After adding any necessary complexity in subsequent iterations, one should then look for empirical fit between simulations and the observable system. Sensitivity analyses should be conducted to help find the simplest model that still captures the processes of interest. Grimm et al.'s (2005) "pattern-oriented modelling" approach emphasizes including key structural elements of the real system that are posited to produce characteristic patterns of that system at multiple scales. If the model is designed to replicate just one empirical pattern, it is often too easy to generate that pattern without the model generating it for the *right* reasons. Replicating multiple patterns mitigates the risk of developing a structurally unrealistic model (Railsback & Grimm, 2012).

One of the most difficult issues of any modelling endeavour, agent-based or otherwise, is finding a desirable level of complexity. If a model is too complex, its analysis may become too difficult or there will be so many parameters that it might be fitted to match existing data even when its structure is an invalid representation of the real system. Conversely, if a model is too simple, it might not capture enough detail to improve understanding of any real system. Pattern-oriented modelling, using patterns at the micro-, meso-, and macro-scale of a system, is a strategy to help guide model development to an appropriate level of complexity (Grimm et al., 2005).

Computational simulation methods, which can include EBM and ABM, are most useful – compared to traditional policy modelling techniques – when there are high levels of uncertainty about a system; but predictions are not reliable under these circumstances (Lempert, 2002; Moss, 2002). Traditional policy analysis might involve defining only a few scenarios. If, on the other hand, we have a large number of relevant future scenarios, we should compare policy options across scenarios and evaluate them according to their "robustness," that is, how well they perform across a wide range of scenarios (Lempert, 2002). ABM and EBM simulations can facilitate this process.

5.2.3. Geographic-based

Geographic information systems (GIS) enable spatial information from a variety of sources to be manipulated in a common projection format such that spatial relationships can be analysed and visualized. Examples of information relevant to climate adaptation for which GIS is an invaluable resource include political boundaries, demographics, infrastructure, weather, response assets, environmentally sensitive areas, hydrology, soil properties, and natural hazards. Government agencies, civil society, and businesses alike, have shifted their practices to collect data in GIS-compatible formats and have created large repositories of relevant spatial information. Moreover, advances in data collection technology are enabling such information to be obtained through a variety of means, including satellites and mobile phones.

Adding to the power of GIS is the ability to store considerable information associated with any point (site) or polygon (area). These attributes provide an opportunity to associate location-specific characteristics with each GIS record. For example, flood mitigation infrastructure that has been assigned geographic coordinates can have information stored behind it that describes its type, age, condition, and use. Similarly, a GIS record for a response asset, such as a disaster aid organization, might contain attributes describing the number of volunteers and their level of training, as well as available supplies. A GIS

demographic layer typically contains attributes characterizing population demographics.

These capabilities relate to climate adaptation policy in a variety of ways. For instance, in order to understand the implications of natural hazards, it is important to consider potential extreme weather event scenarios, superimpose the impact area and intensity of these events, and estimate the loss and damage that may subsequently occur, both immediately and in the long-term. GIS is ideally suited for this purpose (Alam & Mqadi, 2006).

ABMs are useful for representing processes and causal mechanisms underlying, and *generating*, system behaviour. On the other hand, geographic-based models (GBM) represent detailed spatial patterns and facilitate visualization, but are not as adept at representing dynamic heterogeneous processes. Combining ABM and GBM tools allows for a rich understanding of both behavioural *process* and resulting spatial *pattern* (Abdou et al., 2001). For instance, GBM allows us to take into account how changes to spatial representations of an environment – for example, land-use, salinity ingress, and infrastructure systems – might impact agents' opportunities and thus their actions, their interactions with other agents, and the overall system behaviour. For coupled systems, combined GBMs and ABMs are effective tools for integrating disparate types of data (Crooks & Castle, 2012).

5.2.4. Participation-based

Stakeholders will likely view computer-based tools as "black boxes, which raises the issue of their legitimacy and acceptability" (Barreteau & Abrami, 2007). Role-play games (RPGs) may be useful in explaining an ABM or GBM (Barreteau, Bousquet, & Attonaty, 2001). RPGs are used by the Red Cross/Red Crescent Climate Centre to assist researchers and practitioners to better understand climate risks, explore a range of plausible futures, and improve decisions related to mainstreaming climate adaptation into development planning (Mendler de Suarez et al., 2012). Role-play simulations conducted in communities can help to create local climate adaptation plans that have the necessary political momentum to be implemented (Susskind, 2010). RPGs and ABMs have the same conceptual structure (autonomous agents interacting dynamically in a shared environment), allowing them to be combined into a hybrid tool (Barreteau & Abrami, 2007). RPGs can be used to gain understanding of a social system for input into an ABM – information about interactions among actors and their institutions – and to convey an ABM to stakeholders, whereas an ABM can be used to repeat RPGs and explore outside their parameter space (Barreteau et al., 2001).

RPGs, ABMs, and GBMs can be integrated. Knowledge about a physical system is used to populate a GBM and knowledge about a social system is used to characterize agents and their behavioural processes, which together comprise a spatially explicit ABM (Guyot & Honiden, 2006). The RPG – consisting of stakeholders acting with a simplified representation (computational, paper, blackboard, etc.) of the GIS – and the ABM iteratively inform one another, partly by the RPG feeding back into improved behavioural models (Castella, Trung, & Boissau, 2005). The ABM will likely need to be reduced and simplified for conversion into an RPG. Simplifying the ABM and identifying aspects of utmost interest may be done with the assistance of stakeholders (Guyot & Honiden, 2006). Involving stakeholders in co-constructing ABMs and RPGs is often referred to as "companion modelling" (Barreteau et al., 2001). The game is created to observe particular decisions of participants (Castillo, Bousquet, Janssen, Worrapimphong, & Cardenas, 2011). Debriefing sessions – where participants provide feedback on the game – can improve characterizations of agents in the ABM (Bousquet, 2001; Castillo et al., 2011). Stakeholders may not understand the link between their decisions and the larger consequences of those decisions. Participatory modelling can help stakeholders make these connections in an iterative process of describing the environment, their decisions, and running the model (see D'Aquino et al., 2002 for a review of case studies). Developing and validating ABMs with stakeholder input, regardless of whether an RPG is used, can be part of adaptively managing a dynamic, coupled system (Moss, Pahl-wostl, & Downing, 2001).

With the realization that cost–benefit techniques, broadly defined, will often be used for decisions regarding which adaptation projects and policies to implement at scale, we argue for improving the process with the modelling options outlined and community engagement. The process and outputs of formal modelling can illuminate uncertainties and demonstrate the many trade-offs involved under adaptation alternatives. Outputs of models that explore a range of plausible scenarios can inform a cost–benefit analysis, and will ideally incorporate community participation and engagement at both model building and cost–benefit analysis stages.

6. Demonstrating cost-effectiveness and community benefit

The ensuing discussion offers pragmatic context to our review of models and tools, on the understanding that decision-makers share a common goal of ensuring that adaptation interventions seek benefits for the most climate-vulnerable communities. These examples showcase the variability in how decision-support models and approaches are being used by a number of international actors to assess, prioritize, and implement adaptation options at the local level.

There remains significant uncertainty surrounding downscaled climate forecasts, which, along with issues of choosing an appropriate discount rate and valuation technique, complicates any analysis of adaptation projects designed to deliver benefits that may extend decades into the future. The significant uncertainty in the future behaviour of coupled systems regardless of climate uncertainty, further supports our argument for utilizing empirically grounded simulations to better understand change and the potential implications of adaptation options. Cost–benefit analysis, broadly defined, can be an effective, complementary tool for assessing adaptation options and mainstreaming such options into development (Agrawala & Fankhauser, 2008; Stage, 2010).

A range of approaches has proven to be valuable in the assessment of the costs and benefits of proposed adaptation interventions. These include cost–benefit analysis, which has been most widely applied in adaptation cost considerations to date (Berger & Chambwera, 2010), cost-effectiveness analysis, and multi-criteria analysis. The merits of these and other approaches are explored in detail under the Economics of Climate Adaptation working group (2009) and Nairobi Work Programme (UNFCCC, 2011). In contemplating the quantification of costs and benefits of proposed adaptation measures at the local level, it is important to acknowledge several considerations that arise relating to valuation and equity.

Research on the use of cost–benefit analysis for evaluating adaptation in developing countries has highlighted challenges associated with monetizing the costs or benefits associated with issues such as environmental goods and services, social, or cultural values (Chambwera et al., 2011; UNFCCC, 2011). Furthermore, it is important to consider the distribution of costs and benefits, that is, considerations of who benefits from adaptation interventions (UNFCCC, 2011), especially since poorer groups are most vulnerable to climate impacts. Assigning costs and benefits to potential interventions must extend beyond those aspects that can be easily assigned monetary value (such as changes in output of productive systems linked to formal markets) to those that cannot be easily monetized (such as improvements in human well-being and ecosystem services). In local contexts, the social value that disparate groups of individuals place on community assets poses challenges to a traditional cost–benefit approach. For example, in the case of a participatory cost–benefit analysis of drip irrigation in Morocco, non-monetary benefits, including cross-sectoral benefits, were ranked more highly by stakeholders than the monetary benefits (Chambwera et al., 2011). Furthermore, issues relating to the definition of a time horizon and whether to deal with single or multiple baselines are contentious (Chambwera et al., 2011). A decisive variable in most cost–benefit analyses is the discount rate, which, in climate adaptation considerations, Stern (2006) suggests must be lower than in conventional analyses and Broome (2008) believes should be removed from the equation entirely to uphold the principle of inter-generational equity that is central to sustainable development.

Beyond the debates over the variables that contribute to a formal analysis are value judgments that many analyses fail to capture. Of key importance to considerations at the community-level, for example, is failure to capture the distribution of costs and benefits between stakeholder groups (Kennedy, 1981) without using subjective weightings for value judgments (UNFCCC, 2011). These challenges highlight how conventional approaches to assessing costs and benefits must be re-thought in many adaptation-mainstreaming cases. Participatory cost–benefit analysis uses participatory research appraisal methods to ensure all financial, social, and environmental costs and benefits are identified. Piloting this tool in five countries highlighted that not all benefits can be monetized, and it is important not to compare strategies in purely economic terms as this may lead to important benefits being overlooked (Chambwera et al., 2011). An example from Khulna, Bangladesh, showed that the approach can be used to complement quantitative analyses and may even reduce the cost of adaptation by requiring a balancing of benefits across stakeholders (Haque, 2013). The value of adopting a stakeholder-focused approach also lies in facilitating dialogue among stakeholders who may not otherwise interact, as they seek solutions to address their diverse needs (Chambwera et al., 2011). Ultimately, quantitative assessments of costs and benefits should be used not as reductionist simplifications of complex issues, but as decision-support tools in seeking transparent and cost-effective solutions to reducing climate impacts. In considering the pursuit of effective adaptation at scale, it is instructive to examine the extent to which these issues have been reflected to date in decision-making processes in the allocation of international climate finance.

Many adaptation decision frameworks tend to use a form of multi-criteria analysis (MCA) as the basis for decision-making on adaptation strategies, as suggested by UNFCCC (2011). At the adaptation planning level, it is notable that MCA was used to develop National Adaptation Programmes of Action and featured in recent guidance for National Adaptation Plans (NAPs). Some form of MCA is used widely in community-based adaptation; for example, CARE's toolkit suggests prioritization of adaptation strategies based on a set of agreed upon criteria. In selecting a decision-support tool, practitioners must consider the resources required for the analyses, a particularly salient consideration at local levels. For simplicity and ease of use, it is important that the tool is appropriate to the context and purpose.

The entity charged with financing interventions that addresses the needs of the most vulnerable countries and communities is the Adaptation Fund under the Kyoto Protocol. In seeking grant financing from the fund for adaptation interventions in developing countries, proponents must demonstrate that proposals are "cost-effective" and "justified on the full-cost of adaptation reasoning" (Adaptation Fund, 2012). The instructions provided to proponents states that cost-effectiveness is assessed based on a provision of a description of alternative options to the proposed measures and that quantitative assessments of cost-effectiveness are only to be provided where feasible and useful. A review of the technical reviews of proposals (Adaptation Fund, 2013) shows that the assessment of this criterion is undertaken on a qualitative basis in nearly all cases. The remaining criteria applied in the assessment of proposals are predominantly qualitative in nature, with the exception of the assessment of a detailed budget. This practice demonstrates that quantitative modelling and assessment of costs and benefits is not a requirement to obtain funds from the Adaptation Fund, but rather an optional tool that proponents can use to demonstrate that proposed interventions are indeed cost-effective. This flexibility could be perceived as, on the one hand, a lack of support in articulating clear expectations of the review process. On the other hand, in not prescribing the use of quantitative tools, such an approach may benefit community-based adaptation by allowing proponents the flexibility of using the tools best suited to particular local circumstances.

It is expected that the quantitative assessment of costs and benefits will continue to dominate adaptation discussions at the macro-level, where they have proven useful in informing global discourse and choices (Parry et al., 2009; Stern, 2006). At the level of community-based and sub-national interventions, however, issues of valuation, equity, and complexity demonstrate the need for a combined qualitative and quantitative approach, such as the modelling options described herein, to demonstrate how local adaptation needs can most effectively be addressed.

7. Conclusion

Community-based adaptation seeks to incorporate current and future climatic risks into the design of interventions that are key for local economies and overall well-being (Dumaru, 2010; Rojas Blanco, 2006). While communities have extensive knowledge of local environmental changes, they often have limited knowledge of the causes and effects of exogenous change. Building and utilizing integrative models may, in some circumstances, help evaluate and manage trade-offs inherent in local adaptation options. This article has reviewed some tools and techniques available for this purpose.

The uncertainty of projections and lack of understanding of local dynamics means technical data needs to be supplemented with local knowledge (Lunduka et al., 2013). The participation-based models of the type described herein may provide one avenue to achieve this integration.

It is crucial that tools selected for use are appropriate to the situation, remaining cognizant of the resources available for conducting the effort. Under some circumstances, a stakeholder-focused approach to cost–benefit analysis has been deployed, which enables stakeholders to reach an informed consensus based on analyses that take account of both monetary and non-monetary benefits (Lunduka et al., 2013). Whether qualitative or quantitative in nature, however, model and cost–benefit analyses outputs should be seen as decision-support tools rather than as definitive justifications for particular interventions (or for any intervention).

The example illustrating how the Adaptation Fund reviews proposed adaptation options serves to demonstrate how climate finance is attempting to manage these trade-offs and make itself amenable to a variety of approaches and tools. As climate impacts become more severe, it is important that climate adaptation researchers and practitioners, as well as entities charged with the governance of climate finance, maintain this type of flexibility in their analyses and operations to be adaptable themselves to allowing communities to best manage the emerging challenges of climate change and the long-standing challenges of development.

References

Abdou, M. Hamill, L., & Gilbert, N. (2001, October 4–7). *Designing and building an agent-based model in agent-based models of land-use and land-cover change.* Proceedings of an international workshop., Irvine, CA.

Adaptation Fund (2012). *Accessing resources from the adaptation fund [pdf]* Retrieved July 21, 2013 from https://www.adaptation-fund.org/sites/default/files/AdaptationFund%20Handbook%20English.pdf

Adaptation Fund (2013). *Meeting documents.* Retrieved July 30, 2013 from https://www.adaptation-fund.org/afb-meetings

Adger, W.N., Barnett, J., Brown, K., Marshall, N., & O'Brien, K. L. (2012). Cultural dimensions of climate change impacts and adaptation. *Nature Climate Change*, November, 1–6. doi:10.1038/NCLIMATE1666

Adger, W.N., & Kelly, P.M. (1999). Social vulnerability to climate change and the architecture of entitlements. *Mitigation and Adaptation Strategies for Global Change*, 4(3–4), 253–266 (14).

Agrawala, S., & Fankhauser, S. (2008). *Economic aspects of adaptation to climate change: Costs, benefits and policy instruments.* Paris: OECD.

Alam, M., & Mqadi, L. (2006). Designing adaptation projects. *Tiempo*, 60, 21–24.

An, L. (2012). Modeling human decisions in coupled human and natural systems: Review of agent-based models. *Ecological Modelling*, 229, 25–36. doi:10.1016/j.ecolmodel.2011.07.010

Ayers, J. (2011). Resolving the adaptation paradox: Exploring the potential for deliberative adaptation policy-making in Bangladesh. *Global Environmental Politics*, 11(1), 62–88.

Balbi, S., & Giupponi, C. (2009). *Reviewing agent-based modelling of socio-ecosystems: A methodology for the analysis of climate change adaptation and sustainability.* University Ca' Foscari of Venice, Department of Economics Research Paper Series No. 15_09. Available at SSRN: Retrieved from http://ssrn.com/abstract=1457625

Barreteau, O., & Abrami, G. (2007). Combining various time scales through joint use of agent based model and role playing game: The PIEPLUE river basin management game. *Simulation and Gaming*, 38(3), 364–381.

Barreteau, O., Bousquet, F., & Attonaty, J. (2001). Role-playing games for opening the black box of multi-agent systems: Method and lessons of its application to Senegal River Valley irrigated systems. *Journal of Artificial Societies and Social Simulation*, 4(2:5). Advance online publication.

Bassett, T.J., & Zuéli, K.B. (2000). Environmental discourses and the Ivorian Savanna. *Annals of the Association of American Geographers*, 90(1), 67–95.

Berger, R., & Chambwera, M. (2010). *Beyond cost-benefit: developing a complete toolkit for adaptation decisions [pdf]* Retrieved July 30, 2013, from http://pubs.iied.org/pdfs/17081IIED.pdf

Berger, T. (2001). Agent-based spatial models applied to agriculture: A simulation tool for technology diffusion, resource use changes and policy analysis. *Agricultural Economics*, 25(2–3), 245–260.

Bonabeau, E. (2002). Agent-based modeling: methods and techniques for simulating human systems. *Proceedings of the National Academy of Sciences of the United States of America*, 99(Suppl 3), 7280–7287. doi:10.1073/pnas.082080899

Bousquet, F. (2001). Modélisation d'accompagnement – Simulations multi-agents et gestion des ressources naturelles et renouvelables. Mémoire pour l'obtention de l'Habilitation à diriger les recherches. Université Lyon 1.

Bradbury, R. (2002). Futures, predictions, and other foolishness. In M.A. Janssen (Ed.), *Complexity and ecosystem management* (pp. 48–62). Cheltenham: Edward Elgar Publishing.

Broome, J. (2008). The ethics of climate change. *Scientific American*, 00368733, Jun 2008, 298(6), 96–102. doi:10.1038/scientificamerican0608-96

Carley, K. (2002, May 14). Computational organization science: A new frontier. *Proceedings of the National Academy of Sciences*, 99, 7257–7262.

Carmin, J., Dodman, D., & Chu, E. (2013). *Urban climate adaptation and leadership: From conceptual to practical understanding.* OECD regional development working paper 2013/26. Paris, France: Organisation for Economic Co-operation and Development (OECD).

Carter, T.R., Jones, R.N., Lu, X., Bhadwal, S., Conde, C., Mearns, L.O., … Zurek, M.B. (2007). *New assessment methods and the characterisation of future conditions* (pp. 133–171). Fourth Assessment Report of the IPCC, Cambridge University Press, Cambridge, UK.

Castella, J.C., Trung, T.N., & Boissau, S. (2005). Participatory simulation of land-use changes in the northern mountains of Vietnam: The combined use of an agent-based model, a role-playing game, and a geographic information system. *Ecology and Society*, 10(1), 27. Retrieved from http://www.ecologyandsociety.org/vol10/iss1/art27

Castillo, D., Bousquet, F., Janssen, M., Worrapimphong, K., & Cardenas, J. (2011). Context matters to explain field experiments: Results from Colombian and Thai fishing villages. *Ecological Economics*, 70, 1609–1620.

Chambwera, M., Zou, Y., & Boughlala, M. (2011). *Better economics: Supporting adaptation with stakeholder analysis. Briefing IIED: Lessons from adaptation in practice.* Retrieved November, 2011, from http://pubs.iied.org/pdfs/17105IIED.pdf?

75

Cheema, G.S. (2007). *Decentralizing governance: Emerging concepts and practices.* Washington, DC: Brookings Institution Press.

Cioffi-Revilla, C. (2010). A methodology for complex social simulations. *J. Artificial Societies & Social Simulation, 13* (1), 7 http://jasss.soc.surrey.ac.uk/13/1/7.html

Crate, S.A. (2011). Climate and culture: Anthropology in the era of contemporary climate change. *Annual Review of Anthropology, 40*(1), 175–194. doi:10.1146/annurev.anthro. 012809.104925

Crooks, A., & Castle, C. (2012). The integration of agent-based modelling and geographic information for geospatial simulation. In A.J. Heppenstall, A.T. Crooks, L.M. See, & M. Batty (Eds.), *Agent-based models of geographical systems* (pp. 219–251). New York: Springer Science +Business Media B.V.

D'Aquino, P., Barreteau, O., Etienne, M., Boissau, S., Aubert, S., Bousquet, F., … Daré, W. (2002). The role playing games in an ABM participatory modeling process: Outcomes from five different experiments carried out in the last five years. In A.E. Rizzoli & A.J. Jakeman (Eds.), *Proceedings of the international environmental modelling and software society conference* (pp. 275–280). Switzerland: Lugano, 24–27 June.

Dodman, D., & Satterthwaite, D. (2008). Institutional capacity, climate change adaptation and the urban poor. *IDS Bulletin, 39*(4), 67–74.

Dumaru, P. (2010). Community-based adaptation: Enhancing community adaptive capacity in Druadrua Island, Fiji. *Climate Change, 1*(5), 751–763. doi:10.1002/wcc.65

Ebi, K.L. (2009). Facilitating climate justice through community-based adaptation in the health sector. *Environmental Justice, 2* (4), 191–195. doi:10.1089/env.2009.0031

Economics of Climate Adaptation (2009). *Shaping climate resilient development – a framework for decision-making* [*pdf*] Retrieved July 28, 2013, from http://mckinseyonsociety. com/downloads/reports/Economic-Development/ECA_ Shaping_Climate%20Resilent_Development.pdf

Epstein, J.M. (2008). Why model? *Journal of Artificial Societies and Social Simulation*, 11, 4–9.

Filatova, T. (2009). *Land markets from the bottom up. Micro-macro links in economics and implications for coastal risk management.* (PhD thesis) University of Twente.

Fisher-Vanden, K., Wing, I., Lanzi, E., & Popp, D. (2013). Modeling climate change feedbacks and adaptation responses: recent approaches and Shortcomings. *Climatic Change.* doi: 10.1007/s10584-012-0644-9

Gaventa, J., & Barrett, G. (2012). Mapping the outcomes of citizen engagement. *World Development*, 40(12), 2399–2410.

Gilligan, J., Ackerly, B., & Goodbred, S. (2013). "Building resilience to environmental stress in coastal Bangladesh: An integrated social, environmental, and engineering perspective" Bridging the Policy-Action Divide: Challenges and Prospects for Bangladesh, International Bangladesh Development Initiative Conference, Berkeley, California.

Grimm, V., Revilla, E., Berger, U., Jeltsch, F., Mooij, W.M., Railsback, S.F., … DeAngelis, D.L. (2005). Pattern-oriented modeling of agent-based complex systems: Lessons from ecology. *Science* (New York, N.Y.), 310(5750), 987–91. doi:10.1126/science.1116681

Gunderson, L. (2010). Ecological and human community resilience in response to natural disasters. *Ecology and Society, 15*(2), 18.

Guyot, P., & Honiden, S. (2006). Agent-based participatory simulations: Merging multi-agent systems and role-playing games. *Journal of Artificial Societies and Social Simulation*, 9(4). Retrieved from http://jasss.soc.surrey.ac.uk/9/4/8.html

Halsnæs, K., & Traerup, S. (2009). Development and climate change: A mainstreaming approach for assessing economic, social, and environmental impacts of adaptation measures. *Environmental management*, 43(5), 765–78. doi:10.1007/ s00267-009-9273-0

Haque, A.K.E. (2013). *Reducing adaptation costs to climate change through stakeholder-focused project design: The Case of Khulna City in Bangladesh.* Retrieved from http:// pubs.iied.org/pdfs/G03520.pdf

Hertel, T., & Rosch, S. (2010). Climate change, agriculture, and poverty. *Applied Economic Perspectives and Policy*, 32(3), 355–385. doi:10.1093/aepp/ppq016

Hobson, K., & Niemeyer, S. (2011, August). Public responses to climate change: The role of deliberation in building capacity for adaptive action. *Global Environmental Change*, 21(3), 957–971. doi:10.1016/j.gloenvcha.2011.05.001

Holling, C.S. (2001). Understanding the complexity of economic, ecological, and social systems. *Ecosystems*, 4, 390–405.

Huq, S., & Reid, H. (2004). Mainstreaming adaptation in development. *IDS Bulletin*, 35(3), 15–21. doi:10.1111/j.1759-5436. 2004.tb00129.x

Janssen, M.A., & Ostrom, E. (2006). Empirically based, agent-based models. *Ecology and Society*, 11(2). Article no. 37. Retrieved 30 November, 2013 from http://www. ecologyandsociety.org/vol11/iss2/art37

Kennedy, D. (1981). *Cost–benefit analysis of entitlement problems: A critique* [*pdf*] Retrieved July 30, 2013, from http:// duncankennedy.net/documents/Photo%20articles/Cost-Bene fit%20Analysis%20of%20Entitlement%20Problems_A%20 Critique.pdf

Lempert, R. (2002, May 14). Agent-based modeling as organizational and public policy simulators. *Proceedings of the National Academy of Sciences*, 99(suppl. 3), 7195–7196.

Lunduka, R., Bezabih, M., & Chaudhury, A. (2013, January). *Stakeholder-focused cost–benefit analysis in the water sector: Synthesis report.* IIED. Retrieved from http://pubs. iied.org/pdfs/16523IIED.pdf?

Mendler de Suarez, J., Suarez, P., Bachofen, C., Fortugno, N., Goentzel, J., Gonçalves, P., … Virji, H. (2012). *Games for a new climate: Inhabiting the complexity of future risks.* Boston: The Frederick S. Pardee Center for the Study of the Longer-Range Future, Boston University.

Moss, S. (2002, May 14). Policy analysis from first principles. *Proceedings of the National Academy of Sciences*, 99, 7267–7274.

Moss, S., Pahl-wostl, C., & Downing, T. (2001). Agent-based integrated assessment modelling : The example of climate change. *Integrated Assessment*, 2, 17–30.

Nielsen, J.Ø., & Reenberg, A. (2010). Cultural barriers to climate change adaptation: A case study from Northern Burkina Faso. *Global Environmental Change*, 20(1), 142–152. doi:10.1016/ j.gloenvcha.2009.10.002

OECD (2009). *Policy guidance on integrating climate change adaptation into development co-operation.* Paris: OECD.

Olhoff, A., & Schaer, C. (2010). *Screening tools and guidelines to support the mainstreaming of climate change adaptation into development assistance – a stocktaking report.* United Nations Development Programme, New York.

Ostrom, E. (2007). A diagnostic approach for going beyond pana-ceas. *Proceedings of the National Academy of Sciences, 104* (39), 15181–7. doi:10.1073/pnas.0702288104

Ostrom, E. (2009). A general framework for analyzing sustain-ability of social-ecological systems. *Science (New York, N. Y.), 325*(5939), 419–22. doi:10.1126/science.1172133

Parks, B.C., & Roberts, J.T. (2010). Climate change, social theory and justice. *Theory, Culture & Society, 27*(2–3), 134–166. doi:10.1177/0263276409359018

Parry, M., Arnell, N., Berry, P., Dodman, D., Fankhauser, S., Hope, C., … Wheeler, T. (2009). *Assessing the costs of adap-tation to climate change: A review of the UNFCCC and other recent estimates.* Retrieved July 31, 2013, from http://pubs.iied.org/pdfs/11501IIED.pdf

Parunak, H., Savit, R., & Riolo, R. (1998). Agent-based modeling vs. equation-based modeling: A case study and user's guide. *Proceedings of multi-agent systems and agent-based simu-lation. Lecture notes in artificial intelligence* (Vol. 1534, pp. 10–25). Berlin: Springer.

Patt, A., & Siebenhuner, B. (2005). Agent based modeling and adaptation to climate change. *Vierteljahrsheftezur Wirtschaftsforschung, 74*(2), 310–320.

Poteete, A., Janssen, M.A., & Ostrom, E. (2010). *Working together: Collective action, the commons, and multiple methods in prac-tice.* Princeton, NJ: Princeton University Press.

Pringle, P., & Conway, D. (2012). Voices from the frontline: The role of community-generated information in delivering climate adaptation and development objectives at project level. *Climate and Development, 4*(2), 104–113.

Railsback, S., Grimm, V. (2012). *Agent-based and individual-based modeling: A practical introduction.* Princeton, NJ: Princeton University Press.

Robinson, D., Brown, D., Parker, D., Schreinemachers, P., Janssen, M., & Huigen, M. (2007). Comparison of empirical methods for building agent-based models in land use science. *Journal of Land Use Science, 2*(1), 31–55.

Rodima-Taylor, D. (2012). Social innovation and climate adap-tation: Local collective action in diversifying Tanzania. *Applied Geography, 33*, 128–134.

Rojas Blanco, A.V. (2006). Local initiatives and adaptation to climate change. *Disasters, 30*(1), 140–147.

Saito, N. (2013). Mainstreaming climate change adaptation in least developed countries in south and southeast Asia. *Mitigation and Adaptation Strategies for Global Change, 18*(6), 825–849.

Schreinemachers, P., & Berger, T. (2011). An agent-based simu-lation model of human-environment interactions in agricul-tural systems. *Environmental Modelling & Software, 26*, 845–859.

Shaw, R. (2006). Community-based climate change adaptation in Vietnam: Inter-linkages of environment, disaster, and human security. In S. Sonak (Ed.), *Multiple dimension of global environmental change* (Vol. 547, pp. 521–547). New Delhi, India: TERI – The Energy and Resources Institute.

Shepard, P.M., & Corbin-Mark, C. (2009). Climate justice. *Environmental Justice, 2*(4), 163–166.

Smith, J.B., Vogel, J.M., & Cromwell, J.E. (2009). An architecture for government action on adaptation to climate change. An editorial comment. *Climatic Change, 95*(1–2), 53–61. doi:10.1007/s10584-009-9623-1

Stern, N. (2006). *The stern review: The economics of climate change.* London: HM Treasury.

Sterner, T., & Persson, M. (2008). An even sterner review: Introducing relative prices into the discounting debate. *Review of Environmental Economics and Policy, 2*, 61–76.

Sultana, F. (2013). Water, technology, and development: Transformations of development technonatures in changing waterscapes. *Environment and Planning D: Society and Space, 30*(2), 337–353.

Susskind, L. (2010). Responding to the risks posed by climate change: Cities have no choice but to adapt. *Town Planning Review, 81*(3), 217–235.

UNDP/UNEP. (2011). *Mainstreaming adaptation to climate change in development planning. A guidance for prac-titioners.* United Nations Development Programme/United Nations Environment Programme.

UNFCCC (2011). *Assessing the costs and benefits of adaptation options: An overview of approaches. The Nairobi Work Programme and United Nations Framework Convention on Climate Change.* Retrieved July 30, 2013, from http://unfccc.int/resource/docs/publications/pub_nwp_costs_benefits_adaptation.pdf

Wright, H., & Chandani, A. (2014). Gender in scaling up community-based adaptation to climate change. In E.L.F. Schipper, J. Ayers, H. Reid, S. Huq, & A. Rahman (Eds.), *Community based adaptation: Scaling it up* (Chapter 14). New York: Routledge.

VIEWPOINT

Enablers for delivering community-based adaptation at scale

Elizabeth Gogoi[a,b], Mairi Dupar[a,c], Lindsey Jones[a,c], Claudia Martinez[a,d] and Lisa McNamara[a,e]

[a]Climate and Development Knowledge Network (CDKN); [b]Leadership for Environment and Development (LEAD), New Delhi, India; [c]Overseas Development Institute (ODI), London, UK; [d]E3, Bogota, Colombia; [e]SouthSouthNorth, Cape Town, South Africa

1. Introduction

A successful community based adaptation (CBA) initiative which results in the design and delivery of grassroots-generated solutions to adapting to the impacts of climate change is of value to the community involved, but can have little wider impact. If these isolated examples are considered CBA *pilots*, then how to learn from and scale out the pilot is the necessary next phase. CBA pilots can be considered as 'laboratories of innovation'.

There is no single path to delivering CBA at a scale needed to have a significant impact. Mainstreaming CBA pilots into government development plans and programmes is one route, but other actors such as the private sector and NGOs are also capable of delivering CBA at scale. Irrespective of who is the agent, one key challenge is to retain the fundamental principle of community empowerment.

This viewpoint reflects on the work of the Climate Development Knowledge Network (CDKN), a programme funded by the UK and Dutch Governments, which aims to help decision-makers in developing countries design and deliver climate-compatible development. Based on a selection of projects supported around the world it highlights some apparent enabling factors for scaling-out: networks and partnerships; documenting evidence and learning; strong adaptive capacity; and deployment of cost-effective institutional channels and finance mechanisms.

2. Networks and partnerships

In Cartagena, a coastal city in Colombia with a population of only 900,000, the city government, NGOs, and business worked together to assess and understand climate impacts and continue to work towards policy integration, demonstrating how diverse actors' comparative strengths in

technical, financial and human resources and in local knowledge can blend to achieve policy progress.

The city faces immediate and future threats from a changing climate such as flooding and storms, which poses a risk for population displacement and the spread of diseases (Adams, Castrol, Martinez, & Sierra-Correa, 2013). A research institute of marine and coastal research, INVEMAR (Instituto de Investigaciones Marinas yCosteras), forged an alliance in 2011–2012 with the municipality of Cartagena, the Cartagena Chamber of Commerce, and CDKN to undertake the city's first comprehensive and participatory vulnerability assessment and adaptation plans (CDKN, 2013).

The institutional support provided by the municipal government has been central to this achievement. Despite tumult in the mayor's office, with four mayors in two years, there has been constant support by senior civil servants in the city administration. INVEMAR, a well-respected scientific institute, has played a critical role in framing the discussion by, for instance, utilizing down-scaled climate models and graphic imaging to produce artwork showing how the city will be affected by future sea level rise. Despite initial scepticism, various business leaders have also played an important role in ensuring that adaptation options are prioritized within local planning processes.

Scaling-out is occurring simultaneously on three fronts: the geographic area covered by the assessment has expanded to include Cartagena's island territories, as well as the mainland; the national government is watching the Cartagena process closely as it develops and refines national-level adaptation plans; and, other coastal cities in Colombia are looking directly to see how elements of Cartagena's approach can be adopted by their own administrations. Cartagena is seen as a 'first mover' for the leadership and innovation that this city 'community' is showing on adaptation.

Local multi-sectoral partnerships have been the cornerstone to progress in Cartagena so far and will be central to other successful CBA pilots in Colombia and beyond. This catalytic role of 'expert' actors, usually an NGO, research institute or international agency, is typical of most documented CBA pilots. While communities themselves experience climate impacts first hand and are frequently able to deploy indigenous knowledge or generate their own innovations, external experts are in the position to provide resources and technical support and transmit supplementary information. Having a network of such partnerships between the community and outside agents who themselves have national and international connections provides a valuable structure and institutional support for scaled-out CBA.

3. Evidence and learning

Bangladesh, labelled as the 'adaptation capital of the world', has been a pioneer in CBA and has a flourishing NGO sector supporting pilot initiatives throughout the country (Huq, 2013). The country is now establishing platforms and networks that are bringing visibility and accountability to CBA pilots in order to encourage scaling-out. Over 20 INGOs and research partners are implementing a programme of research called 'Action Research on CBA in Bangladesh' (ARCAB) as well as providing a platform to jointly promote CBA in Bangladesh and beyond. An annual international conference on CBA is now a regular fixture and this emerging 'community of practice' has also spread to two new online forums (weADAPT and the CBA Exchange) further increasing the visibility of CBA initiatives. While there is anecdotal evidence that the exposure provided to CBA pilots and practitioners through these networking opportunities has resulted in some instances of scaling-out, research is required to confirm this (Gundel, Anderson, Kaur, & Schoch, 2013).

A visible CBA pilot will only be taken further and considered for scaling-out by an institution if it can prove its value. ARCAB has recognized this and developed a framework, tool and manual for participatory monitoring and evaluation (M&E) of CBA which considers whether achievements match expectations, whether achievements were the right ones, and whether CBA is being done in the right way and reaching the right scale (Action Research for Community Adaptation in Bangladesh, 2012). For example, by using this approach ActionAid found that while their CBA projects in Bangladesh have moved beyond business-as-usual approaches to development and disaster risk reduction, the communities need further support for anticipating and adapting to longer term climate change risk (Faulkner & Ali, 2012). M&E also needs to go beyond looking at how effective the CBA pilot is, to how efficient it is and the scale of benefits for communities relative to the level of investment (Sova, Chaudhury, Helfgott, & Corner-Dolloff, 2012). One of the key reasons why there have been so few success stories for scaling-out is that they are considered too time- and resource-intensive.

The work of ARCAB and others in Bangladesh is testing the hypothesis that a CBA pilot needs to be visible, well-documented and assessed in order to instigate scaling-out by other institutions and actors. Their efforts are illustrating the connection between understanding and learning from CBA pilots and scaling-out.

4. Ensuring CBA supports characteristics of adaptive capacity

A core challenge in scaling-out CBA beyond a piloting stage is the localized nature of adaptive capacity and opportunities for delivering effective adaptation. Success factors in one community may not translate or be replicable in another community context. Despite this, broad commonalities can be drawn from across a range of contexts which help one to focus attention on how CBA pilots can be scaled out.

The Africa Climate Change Resilience Alliance (ACCRA), a network of five INGOs and research partners working in Mozambique, Ethiopia and Uganda, is attempting to better understand these commonalities. Their Local Adaptive Capacity (LAC) framework breaks the adaptive capacity into five characteristics: access to and availability of assets; institutions and entitlements; knowledge and information; innovation; and flexible forward looking decision-making (see Jones, Ludi, & Levine, 2010). ACCRA used the LAC to assess how community-level projects impact on each of the characteristics of adaptive capacity. Their findings have important implications for the design and delivery of CBA at scale.

First, the characteristics of adaptive capacity do not act in isolation: they interact and depend on each other. Therefore, community-level adaptation interventions that focus simply on a single characteristic – such as the provision of physical assets and capitals – are unlikely to address the full spectrum of processes needed to support adaptive capacity (Levine, Ludi, & Jones, 2011). Successful adaptation happens when people understand and fully exploit the interconnected nature of each characteristic. Second is a need to 'rethink' participation; for community priorities to be adequately recognized in the implementation of CBA requires meaningful engagements with local actors, addressing power imbalances, and a two-way sharing of knowledge and information that is rarely achieved in the delivery of 'traditional' development objectives (Levine et al., 2011). Third, greater emphasis is needed on supporting autonomous innovation. People's own ability and practice of experimentation and innovation is one of the key manifestations of their agency. For CBA activities to successfully capitalize on this requires (a) an understanding of how local agents are innovating and (b) an understanding of the constraints to experimentation and uptake of new ideas.

ACCRA's experiences serve to document how understanding the factors that support a community's capacity to adapt to change is crucial for successfully scaling-out CBA. More importantly, the network's findings show how CBA activities can learn from the experiences – successes and failures – of wider development interventions, many of which are well documented.

5. Finance and institutional mechanisms

An institution needs to be willing and able to deliver CBA at scale. The government would be the obvious choice, with political will and competent and decentralized institutions and governance appearing to be necessary conditions. In fact, the decentralization of appropriate levels of decision-making and management has to go further and rest with the community itself if CBA is to retain its unique community-driven character. The institutional and funding structures to manage scaling out a CBA pilot has to be flexible, and allow culture and indigenous norms or knowledge to determine outcomes.

Private sector actors can in some instances be the agent for scaling-out CBA pilots. For example, Dunavant Zambia Ltd. is the largest cotton ginning company in Zambia, working with over 100,000 small-holder farmers annually through contract farming systems. This involves prior agreements for farmers to produce an agricultural product in a certain manner for a buyer for a guaranteed price. Dunavant Zambia has recognized the potential of community-based agroforestry and conservation farming programmes for improving yields, which have been pioneered in the country since the mid-1990s by Zambia's Conservation Farming Unit and the World Agroforestry Centre (Ward, 2011).

Although many climate-resilient agricultural technologies such as tree crops and conservation farming practices yield long-term benefits, their high upfront input costs discourage adoption by small-holder farmers (Innovations for Poverty Africa [IPA], 2013). However, DunavantZambia have a commercial interest in encouraging soil fertility and higher future agricultural yields. They, therefore, formed a partnership with the NGO Shared Value Africa through the 'Trees on Farms' programme to encourage the adoption of fertilizer trees by their farmers as a strategy to increase long-term soil fertility and crop yields.

The scaling-out process for the Trees on Farms programme has not yet reached the expected level, but shows positive signs. In November 2011, around 2500 farmers received training on tree planting, and there has been high demand for the programme, with 83% take-up for the 1300 farmers studied (IPA, 2013). At the end of the first year, there were a total of 19,400 surviving trees under the care of 700 farmers in the research group (IPA, 2013). DunavantZambia has plans to scale up the current programme to the central and southern parts of the

country to reach 10,000 or more farmers in 2013 (K. Jack, interview, May 14, 2013).

This case indicates that the commercial interests of companies and the climate change and development objectives of communities and NGOs can positively coincide and be a strategy for scaling-out CBA pilots. Private sector flexibility and openness to innovation make commercial companies important role-players in experimenting with novel ways to expand to new communities (K. Jack, interview, May 14, 2013). Commodity crop firms such as DunavantZambia also have existing systems and infrastructure such as training services for farmers that can be harnessed for CBA, which may improve CBA efficiency and cost-effectiveness. However, there are challenges to using the private sector to deliver CBA such as the need to protect the participatory nature of CBA which is focused on the most vulnerable. Government will also remain an important actor by providing supportive policy frameworks, such as subsidizing natural fertilization methods (Ward, 2011).

6. Conclusion

These lessons from Africa, Asia and Latin America on scaling-out CBA suggest there is not one single model or path to scaling out of CBA pilots. Mainstreaming within the development planning process is just one option, with other actors such as the private sector also being potential delivery partners. The context and specifics determine where and how scaling-out can be achieved. However, CDKN's experience suggests that there are some key enabling factors which support the process. Further documentation and assessment of instances when the impact of an individual CBA pilot has been expanded to a larger scale are required. A more rigorous analysis of these enabling factors would help practitioners design and implement CBA pilots, and help external actors, including governments and donors to provide the correct enabling environment.

References

Action Research for Community Adaptation in Bangladesh. (2012). *ARCAB M&E and baseline strategy for CBA: Final report.* November 2012.

Adams, P., Castrol, J., Martinez, C., & Sierra-Correa, P. (2013). *Embedding climate change resilience in coastal city planning: Early lessons from cartagena deIndias.* London: CDKN.

Faulkner, L., & Ali, I. (2012). *Movingtowards transformed resilience: Assessing community-based adaptation in Bangladesh.* Dhaka: ActionAid Bangladesh, Action Research for Community Adaptation in Bangladesh, International Centre for Climate Change and Development, and Bangladesh Centre for Advanced Studies.

Gundel, S., Anderson, S., Kaur, N., & Schoch, C. (2013). *Assessing the CBA community of practice?* London: IIED Briefing Note.

Huq, S. (2013). Climate change experts head to 'adaptation capital of the world', IIED Press Release. Retrieved June 5, 2013, from http://www.iied.org/climate-change-experts-head-adaptation-capital-world

Innovations for Poverty Africa. (2013). Encouraging the adoption of agroforestry: A case study in Eastern Province, Zambia. Practical lessons learnt. Retrieved May 14, 2013, http://www.poverty-action.org/Lusaka_Chipata_Agroforestry

Jones, L., Ludi, E., & Levine, S. (2010). *Towards a characterisation of adaptive capacity: A framework for analysing adaptive capacity at the local level.* London: Overseas Development Institute.

Levine, S., Ludi, E., & Jones, L. (2011). *Rethinking support for adaptive capacity to climate change. The role of*

development interventions. London: Overseas Development Institute.

Sova, C. A., Chaudhury, A. S., Helfgott, A., & Corner-Dolloff, C. (2012). *Community-based adaptation costing: An integrated framework for the participatory costing of community-based adaptations to climate change in agriculture* (CGIAR Research Program on Climate Change, Agriculture and Food Security (CCAFS) Working Paper No. 16). Colombia. Retrieved from www.ccafs.cgiar.org

Ward, M. (2011). *The case for evergreen agriculture in Africa: Enhancing food security with climate change adaptation and mitigation in Zambia.* London: CDKN.

REVIEW ARTICLE

Mainstreaming children's vulnerabilities and capacities into community-based adaptation to enhance impact

Paul Mitchell[a] and Caroline Borchard[b]

[a]Save the Children, Hanoi, Vietnam; [b]Plan International, Bangkok, Thailand

Children are particularly vulnerable to the impacts of climate change. Despite this, relatively little attention is paid to building their capacity to manage the impacts of climate change they experience now and those they will experience in future. While child-centred approaches are starting to emerge in the field of community-based adaptation, these approaches are almost exclusively used by child-focused organizations. This article argues that mainstreaming children's needs and capacities into broader adaptation efforts can lead to more sustainable outcomes that can help to build long-term community-level adaptive capacity. A series of short examples from the field are used to highlight the different contexts in which child-centred approaches to community-based adaptation are taking place and some outcomes achieved to date. The article concludes that while there is a growing body of anecdotal evidence that taking a child-centred approach to community-based adaptation can build the adaptive capacity of children and also provide benefits to entire communities, there is no solid evidence-base proving that what has worked in a growing number of cases is more broadly applicable, translatable to other regions or sustainable in the absence of direct project support. The article recommends that collaborative efforts between researchers and practitioners should be launched to gather this evidence.

Introduction

Children are widely considered as one of the groups most vulnerable to the impacts of climate change (Bartlett, 2008; Intergovernmental Panel on Climate Change [IPCC], 2007c, p. 52, 2012; Save the Children, 2009; Seballos, Tanner, Tarazona, & Gallegos, 2011; United Nations Children's Fund [UNICEF], 2011). It is today's children who will experience a significant intensification of climate-related disasters and, consequently, a key focus of soft actions for adaptation should be directed to building the adaptive capacity of children now in order to ensure that children's rights (as children and future adults) are protected in a changing climate.

This article provides an overview of the key climate-vulnerabilities facing children and describes the nascent child-centred approach to community-based adaptation, arguing that mainstreaming children's needs and capacities into broader adaptation efforts can lead to more sustainable outcomes and can help to build long-term community-level adaptive capacity. The article also argues that integrating a climate risk approach into broader child-centred community development programming can help to make the impact of sector-based programmes more sustainable in a changing climate. A series of short examples from the field are used to highlight the different contexts in which child-centred approaches to community-based adaptation are taking place and some outcomes achieved to date. The article concludes that while there is a growing body of anecdotal evidence that taking a child-centred approach to community-based adaptation can build the adaptive capacity of children and also provide benefits to entire communities, there is no solid evidence-base proving that what has worked in a growing number of cases is more broadly applicable, translatable to other regions or sustainable in the absence of direct project support. The article then recommends that collaborative efforts between researchers and practitioners should be launched to gather this evidence.

Climate change and children

The Convention on the Rights of the Child (CRC) defines children as 'every human being below the age of 18 years' (UN [United Nations], 1989, art.1).[1] It is widely acknowledged that children, as a broad group, are particularly vulnerable to the impacts of climate change for reasons related to their physical and mental development as well as their general exclusion from decision-making processes (Bartlett, 2008; IPCC, 2007c, p. 52, 2012; Save the Children, 2009; Seballos et al., 2011; UNICEF, 2011). Children are affected by nearly all the conceivable impacts of climate change. The ways in which communities and governments plan for and respond to the unavoidable impacts of climate change through policy processes and practical actions will, in large part, shape the ways in which children experience a climate-changed future.

Climate change holds the capacity to stall and even reverse gains in development achieved in recent decades and, accordingly, to impact on the full range of children's rights enshrined in the CRC. States Parties to the CRC have committed to ensuring a broad range of children's rights – including rights to survival, development, protection and participation – are safeguarded through national policy and legislation. Despite near universal endorsement of the CRC, many children are not even aware they have rights, or, if they are, have little ability to claim them. This existing 'rights deficit' affects children's development and increases their vulnerability to climate change impacts.

Why are children vulnerable?

Physiological impacts

The most obvious reason children are more vulnerable than (most) adults to climate change is physiological. 'Children are especially sensitive to changes in the climate because they are physiologically and metabolically less able than adults at adapting to heat and other climate-related exposure' (UNICEF, 2011, p. 1. See also Akachi, Goodman, & Parker, 2009, p. 2). Nearly every impact of climate change in every region holds the potential of significant impacts on children. Children are at greater risk of being injured or killed in a disaster (Bartlett, 2008; Costello et al., 2009; Telford, Cosgrave, & Houghton, 2006; UNICEF, 2007, p. 6; Waterson, 2006), the majority of which are weather-related. Children are also more susceptible to heat stress and the health impacts of increased diarrhoeal and respiratory disease and changing ranges of vector-borne diseases, which we will likely see as our climate continues to change. They are also more at risk from more gradual impacts like under-nutrition and malnourishment as food insecurity increases (see Costello et al., 2009; Sheffield & Landrigan, 2011, p. 291). Moreover, impacts can have much longer term impacts on the life of a child as certain incidences, such as malnutrition or illness during the phase of rapid physical and mental development can have long-term effects on physical and mental capabilities (Bartlett, 2008).

Psychological impacts

While children's bodies are more easily damaged by extreme weather and other impacts of climate change, they are also generally more likely to experience psychological trauma associated with climate change (see Farrant, Armstrong, & Albrecht, 2012; Frumkin, Hess, Luber, Malilay, & McGeehin, 2008, p. 440). Doherty and Clayton (2011, p. 265) distinguish three levels of psychological impacts of climate change: acute and direct impacts (e.g. trauma from directly experiencing extreme weather events or the loss family members during a disaster); indirect and vicarious impacts (e.g. intense emotions associated with observation of climate change effects at broader scales) and psychosocial impacts (e.g. of violence over increasingly scarce resources and trauma of climate-induced migration). As the impacts of climate change intensify over coming decades, more and more children are likely to be exposed to one or more of these psychological impacts, with poor children least likely to be shielded from damage and less likely to be in a position to seek or receive adequate assistance after any climate change-related event. The IPCC's (2007b, p. 399) Fourth Assessment Report concurs, stating that 'There is also evidence of medium to long-term impacts [of climate change] on behavioural disorders in young children.' These psychological impacts are likely to have long-term implications for children's ability to live a productive life – particularly children in poor communities. However, longitudinal studies will be required to definitively prove this link.

Exclusion

The exclusion of children from decision-making processes is a common practice throughout the world and is contrary to Article 12 of the CRC, which provides:

> States Parties shall assure to the child who is capable of forming his or her own views the right to express those views freely in all matters affecting the child, the views of the child being given due weight in accordance with the age and maturity of the child. (UN, 1989, art12)

While Article 12 should not be interpreted as giving children complete autonomy in decision-making processes, 'it does introduce a radical and profound challenge to traditional attitudes, which assume that children should be seen and not heard' (Lansdown, 2001, p. 2). This denial of children's agency and right to participate in decision-making processes that directly affect their lives has implications for whether these processes will adequately respond to their needs. It also limits the extent to which these processes can benefit from the unique insights and innovative solutions that children can generate. Seballos et al. (2011, p. 39) found that adults' perceptions of children's agency is a crucial factor in deciding whether children are able to effectively engage in disaster risk reduction (DRR) activities. This finding is highly likely to be echoed in children's engagement in community-based climate change adaptation actions, given the significant overlaps between community-level responses to disasters and climate change, and the DRR and climate change communities' similar modes of operation.

Child-centred approaches to adaptation

While community-based responses to climate change are growing at an exponential rate, children's active participation in adaptation planning and implementation remains limited. Despite the still small number of projects specifically focusing on (or factoring in) children's unique vulnerabilities and adaptation needs, research is starting to show that children have much to contribute to building community-level adaptive capacity than may be readily apparent (see, e.g. Tanner, 2010; Tanner et al., 2009). A recent study by UNICEF and Plan International found that:

> The evidence [...] suggests that the benefits of child-focused approaches to [climate change] adaptation are likely to be high – because children are numerous and experience the impacts of climate change more acutely than other groups and over a longer period, the avoided losses associated with adaptation to both sudden disasters and systemic climate change are significant. (UNICEF & Plan International, n.d., p. 24)

Taking a child-centred approach to development and humanitarian action will necessarily help organizations focus on the specific vulnerabilities of children. The child-centred approach to development is grounded on the recognition of child rights and places children (and their caregivers) at the heart of efforts to secure their rights and fulfil their development aspirations. Anecdotal evidence (some of which is outlined below) is starting to show that using the child-centred approach as an entry point to build broader community understanding of, and action on, climate change can be very effective. While approaches to adaptation do not necessarily have to be child-*focused* or child-*led* in their entirety to effectively address the needs and capacities of children, taking a child-*centred* approach to elements of projects working on issues relevant to children will ensure their needs and capacities are addressed in ways that broader approaches may not. In fact, experience in the field at project level (as outlined below) is leading the authors, as community-based adaptation practitioners, to ascertain that mainstreaming child-centred approaches to working with children and with their caregivers and decision-makers into broader sector- or issue-based approaches to adaptation can help address multiple barriers to effective adaptation (including those based on race, class and gender, as well as age) without the need to launch specific

separate adaptation projects running in parallel to other community projects (which can cause confusion and generate inefficiencies).

Child-centred approaches to adaptation are not just about increasing the participation of children in the decision-making forums that affect their lives – though this is a crucial element of ensuring adaptation processes do effectively address the needs and capacities of children – it is 'equally about engagement with support structures and institutions, including households, communities, local and national governments, and international organisations, to minimise adverse impacts and reduce or mitigate the risks that directly affect children's lives' (Save the Children, 2013a, p. 3). Increasing the understanding and ability of caregivers (parents, teachers, community leaders, local and national government) to include a focus on the children's needs and capacities in a systematic way when addressing current and future climate change risks (whether or not children themselves are in the room) is key to securing children's rights in a climate-changed world.

Mainstreaming child-centred approaches into community-based adaptation – key considerations

Mainstreaming – as regards to addressing climate change risks and adaptation through or within development policies and programmes – as a concept, 'has been borrowed from the development discourse' (Klein, Schipper, & Dessai, 2003, p. 8), where it has been most successful with regards to gender, at the level of policy development at least (Moser & Moser, 2005, p. 12). There are a range of definitions of mainstreaming, but most include the key elements of integrating policies and/or measures to address climate change risks through development planning and/or sector-based decision-making processes (see, e.g. Dalal-Clayton & Bass, 2009, pp. 19–21; Huq & Reid, 2004, pp. 19–20; Schipper 2007, p. 7; United Nations Development Programme [UNDP] & United Nations Environment Programme, 2011). For the purposes of this article, we take mainstreaming of child-centred approaches to adaptation to mean *both* including the children's needs and capacities into the planning and implementation of adaptation measures *and* increasing the voice and agency of children in the adaptation decision-making processes that affect them.

There are two key considerations that need to be taken into account when mainstreaming child-centred approaches into sector- or issue-based adaptation projects: that age alone is not the most useful indicator of vulnerability; and that capacity provides a better entry point for building resilience.

1. *Age alone is not a useful indicator of vulnerability*

'Vulnerability is related to predisposition, susceptibilities, fragilities, weaknesses, deficiencies or lack of capacities that favour adverse effects on the exposed elements' (IPCC, 2012, p. 70). While children are more vulnerable than other groups to a variety of climate change impacts, not all children are equally vulnerable, nor are all children more vulnerable than all adults. As Bartlett (2008, p. 509) points out, 'some children may actually be more resilient than their elders'. There is a risk that treating children as a homogenous group will serve to mask or erase other factors that enhance or reduce vulnerabilities (including poverty, class, disability and power relationships). The category 'children' includes a broad membership – from birth to age 18. It is, thus, necessary to take into account the different capacities, capabilities and vulnerabilities of different subgroups and even individuals.

Children are generally aggregated into a broad 'vulnerable' group in policy and strategy documents from international to local levels, usually with women (another broad non-homogenous group), elderly, disabled and sometimes indigenous peoples (IPCC, 2012, p. 70). Such blanket descriptions of 'vulnerable' groups usually fail to give an indication as to what these groups are vulnerable to (IPCC, 2012) and provide differentiation of levels of vulnerabilities within these groups (e.g., vulnerabilities of children under-five compared to vulnerabilities of adolescents; or girl child versus boy child). This results in a generalized failure to meet the needs of any one subgroup and serves to erase differences within groups. Girls, for example, often face particular vulnerabilities during climate-related disasters and barriers to adaptation due to their domestic roles and responsibilities and lower access to education (Plan International, 2011).

Hence, it is important to take the time and effort to break down differential vulnerabilities by age, gender and status in community-led vulnerability and capacity assessments. Children have an important role to play in this assessment process as they have their own knowledge of hazards, hazardous places and vulnerability that is often different than adults (Gaillard & Pangilinan, 2010; Plush, 2009). Despite this, the concept of 'children' as a group, limited by age (birth to age 18), can be a useful tool for advocacy and policy engagement. While homogenization is risky when implementing child-centred approaches to adaptation, when arguing for greater involvement of children in climate change-related decision-making at all levels it is, in fact, necessary. Given the choice between getting children on the agenda broadly *or* increasing understanding of the differences within and between subgroups within the grouping of *children*, the former needs to be used as a tool to pave the way for the latter. Once the concept of children's vulnerability – broadly understood and even homogenized – becomes a key part of local, national and international climate change planning, child-centred approaches can then work to increase understanding of, and capacity to respond to, the different capacities and vulnerabilities that exist between genders, ages and socio-economic status subgroups within the category of children.

2. *Capacity provides a better entry point for building adaptive capacity*

While it is important and useful to understand children's specific vulnerabilities to climate change, if we are to put their rights at the centre of adaptation actions, we must also understand (and value) their capacity to create change in their communities. Adaptive capacity is more than merely the absence of vulnerability and we know that, despite their inherent vulnerabilities, children can be extraordinarily adaptive in the face of stresses and shocks, especially if they are actively involved in responses to them (Bartlett, 2008; IPCC, 2012). Adaptive capacity is a very broad concept, which the Intergovernmental Panel on Climate Change (IPCC, 2007a, 17.3.1) defines as 'the ability or potential of a system to respond successfully to climate variability and change, and includes adjustments in both behaviour and in resources and technologies'. Adaptive capacity is also highly context-specific (Smit & Wandel, 2006, pp. 286–287). A full treatment of adaptive capacity in children is beyond the scope of this article. Rather, this article builds on the examples from the field, provided below, to focus on the contribution that children's knowledge, awareness and participation make to increasing adaptive capacity (theirs and their communities').

Moreover, children's active engagement in community-based action to reduce climate and disaster risks has benefits not only for the development of their own adaptive capacity, but it can also be a source of energy, resourcefulness and knowledge for broader community-based adaptation efforts (Bartlett, 2008; IPCC, 2012, p. 314; Tanner, 2010). As children interact with other children and adults, if they are well informed and supported, they can be effective channels of information, role models and agents of change. Also, by developing children's understanding of risk and ways to manage it, interventions are more likely to have a sustainable impact in the medium to long term (Turnbull, Sterrett, & Hilleboe, 2013, p. 19).

In recent years, child-focused organizations have tailored the child-centred approach to development for use in community-level adaptation activities that take as their starting point that a focus on building children's adaptive capacity will increase a given community's ability to manage the impacts of climate change – particularly in the longer term. This approach generally works across sectors and disciplines, helping community development projects address risk at multiple levels:

> It is an approach that not only addresses system-level interactions (economic, environmental, political, social) and how they may aggravate current and future risks; but also

focuses on the individual – ensuring they not only have the necessary tools to minimise the impacts of shocks and stresses, but are also capable of adapting to new realities and changing contexts. (Save the Children, 2013a, pp. 3–4)

Putting children at the centre of actions to build adaptive capacity can have unexpected and innovative results. Examples from recent projects implemented at the community level by child-centred non-government organizations (NGOs) in a range of contexts can serve to highlight some of the broad range of outcomes that a child-centred approach to adaptation can achieve.

Child-centred approaches to adaptation: lessons from developing countries

A range of child-centred organizations are now developing and implementing child-centred approaches to community-based adaptation. Plan International and Save the Children International, two significant global child-centred NGOs, are increasingly engaging in climate change as a development issue and working at the community level, through child-centred approaches to adaptation, to build the adaptive capacity of children and their communities. Many of these activities take place in the education sector and focus on increasing children's and caregivers' understanding of climate change impacts, local vulnerabilities and adaptive capacities, and how to develop and implement adaptation plans.

The examples highlighted below are drawn from recent activities supported by Plan and Save the Children and are all anchored, one way or another, in the education sector (e.g., because they may take place in school contexts, or because they may include a significant focus on increasing understanding to catalyse action). The examples below serve as a sample of recent approaches to integrating child-centred approaches to adaptation into the education sector (broadly defined) and highlight that working through existing systems and structures focused on children (like schools and children's clubs) can enhance adaptation outcomes.

Climate change and education

In many instances, education can effectively provide the appropriate knowledge, skills and behaviour change that successful climate change adaptation requires – as highlighted under Article 6 of the United Nations Framework Convention on Climate Change, which includes specific reference to the need to ensure 'public participation in addressing climate change and its effects and developing adequate responses' (UN, 1992, p. 17). Increased knowledge and awareness of the likely impacts of climate change at local scales and locally appropriate adaptation options enables individuals and communities to make informed decisions about how to adapt individual lives and livelihoods as well as take action for climate-resilient sustainable development (Anderson, 2010).

While recognizing the importance of including specific content in the curriculum, Bangay and Blum (2010) argue that the challenges of climate change require all concerned to examine the degree to which existing educational provision is adapted to and preparing people for radically different futures. There is thus a renewed emphasis on teaching and learning methodologies that are participatory, experiential, critical and inclusive with a focus on building essential life skills for a future in a rapidly changing world (Anderson, 2010). Climate change education needs to engage the full range of educational channels – formal and non-formal, and from primary through tertiary and adult education (Bangay & Blum, 2010). But it is also critical to provide opportunities for children to apply their newly gained knowledge to better facilitate change in communities.

Creating tomorrow's climate activists – building knowledge and catalysing action

A lack of information is often the first barrier to action but this is sometimes easily overcome. When given access to information, young people can be very quick to grasp the implications of climate change and also highly capable of advocating for change in their communities and beyond (Institute of Development Studies, 2009).

At a 2013 workshop in the mountainous-coastal rural province of Aurora in the Philippines, students from schools in two districts spent time learning about climate change and what it means for their lives. The children learned about their rights (as enshrined in the CRC) and how climate change will make it more difficult for them to ensure their rights are fulfilled.

Climate change is fundamentally a rights issue, especially for today's children whose human rights (both now and in future) are at the greatest risk of violation in a harsher climate. In Aurora, the students identified the rights they considered most at risk in a changing climate – those that mattered most to them:

- The right to play and rest (CRC article 31);
- The right to health care, clean water, food and a clean environment (CRC article 24);
- The right to an education that develops talent, personality and ability (CRC article 29);
- The right to food, clothing, a safe place to live and to have basic needs met (CRC article 27);
- The right to protection from any kind of exploitation (CRC article 36);
- The right to live with a family who cares for you (CRC article 9).

They went on to develop advocacy messages for different audiences (peers, caregivers, policy-makers) and to work out which mediums might work best for which audiences (including video, radio, comic strips, theatre, lectures, stories, and demonstrations, for example). Armed with new knowledge and advocacy messages and mechanisms, the children are now developing school level strategies for increasing understanding and catalysing action, supported by small grants with which to test the concrete adaptation actions agreed at the school level (Save the Children, 2013b).

Education and awareness raising as a means to catalyse action through advocacy has also recently proved effective in a small project in Papua New Guinea, where the leaders of a community had been attempting to get government support for the construction of a seawall as an adaptation solution to local erosion and storm surge issues. Children participating in a community-based adaptation project supported by Plan International and the Foundation of the Peoples of the South Pacific produced a poster and short video highlighting the impact of coastal erosion on their community which was shown to a Member of Parliament, who immediately secured funds to commence seawall construction (Foundation of the Peoples of the South Pacific International & Plan International, 2013).

Institutionalizing risk reduction education – getting DRR and climate change adaptation into schools

Increasing knowledge and understanding among relatively small groups of children is certainly useful, but for change to be sustainable it needs to be systemic. It is notoriously difficult to get "niche" issues into national curriculums. Many countries already feel their curriculum is full and many children probably agree – but this is likely to have as much to do with modes of teaching as the number of topics covered. Civil society and NGOs have, for a number of years, been implementing DRR learning activities within school structures, but outside the curriculum – through parallel activities like DRR clubs often operating on school grounds but out of school hours. Others are even further removed, operating out of children's clubs in community centres rather than on campus. However, increasingly governments have made commitments to integrate climate change and DRR into education as manifested in Article 6 of the United Nations Framework Convention on Climate Change (UN, 1992). In the global education community, several stakeholders, including United Nations agencies (UNESCO, UNEP, and UNICEF), SEAMEO – the Southeast Asian Ministers of Education grouping, and NGOs (ActionAid, Plan International, and Save the Children International), are supporting national efforts to incorporate components of the climate change and DRR agenda in education systems and helping

schools and communities build the skills of communities and learners to adapt to climate change (Anderson, 2010).

In Vanuatu, the government recently initiated a process of national curriculum reform, which, when finalized, will include "cross-curricular components" common to all grade levels. Disasters and climate change are, by their nature, cross cutting issues affecting all sectors and all people. Teaching climate change only in science classes (as is the case in many curriculums) risks isolating it from the human processes that drive it and the human impacts it inflicts. Including concepts and information on disaster and climate risks and resilience strategies within the "cross-curricular components" would reduce the likelihood of reducing climate change to a scientific or environmental issue and ensure it is communicated across the broad range of subjects upon which it impacts. In the first phase, curriculum materials and teachers' guides have been developed and piloted in several primary grades with support from NGOs. A preliminary evaluation of the teaching materials and learning outcomes has indicated that integrating disaster and climate risks and resilience into the curriculum has been an effective way of improving children's understanding of hazards and disasters and, importantly, their knowledge about how to keep themselves and their communities safe. This gives a strong impetus for the next phase of the project, which will formally include the materials in the new national curriculum (Save the Children, 2013c).

In Vietnam, this process is further advanced. The Ministry of Education and Training and NGOs have jointly produced a booklet on climate change for students and a reference guidebook for teachers and facilitators. These resources were approved by the Ministry of Education and Training in mid-2012 and have begun initial rollout to schools supported by NGO activities. To date, over 300 primary and secondary level teachers from 57 schools in 5 provinces have received training to utilize the materials (UNEARTH News, 2013).

While significant challenges remain in integrating climate change into education systems, these are encouraging signs that the next generation of leaders will be better equipped to manage its impacts.

DRR goes viral – self-replication of risk reduction education

If national adoption of school-based risk reduction education is not feasible, there is always the option to go viral. In a project in Timor-Leste, NGOs were working with two schools in a remote province to help children gain a better understanding of the risks posed by current climate variability and extremes. Once the school communities learned about the risks they and their communities face from the already harsh climate (without even explicitly considering the future impacts of climate change) they decided that other schools in the area needed access to the same information and set about providing it. Through the efforts of teachers the reach of the project was more than doubled, with over 12,000 children in 118 schools gaining access to information and resources with which to influence community planning processes (Save the Children, 2012).

While this process of sharing educational materials and new knowledge was entirely person- and paper-based, it is easy to see how, in this age of mobile connectivity, messages and educational materials could be circulated more widely at a greater speed through simple text message based systems, similar to those used for disaster preparedness and early warning systems in many countries in the Global South. While climatic change is an inherently complex process, its implications (at the level of well understood trends) are fairly simple to grasp – as are the far reaching implications of inaction. Communicating easily understood messages via mobile phone could provide a useful extension of broader school and community-based climate change education activities. For example in the Philippines, children trained about climate-smart DRR decided to organize a "DRR Texter Clan". Using mobile phones, they sent text messages to their friends and a wider youth network about weather warnings, tips for disaster preparedness and risk reduction, and raised awareness about child protection in emergencies (United Nations Office for Disaster Risk Reduction & Plan International, 2012, pp. 15–16).

New approaches for urban contexts

As the trend of rapid urbanization continues, there is an increasing need to work with urban children across the full range of development sectors. Climate change poses significant risks to urban children, although the type of risks and vulnerabilities might differ from their rural counterparts. Organizations facilitating community-based adaptation draw heavily on participatory methodologies that were developed for community-based DRR in rural agricultural communities, such as hazard ranking, seasonal calendars, transect walks and community mapping, and vulnerability and capacity assessments. However these tools for analysis may be less directly applicable in urban areas. In Bangladesh, children in Mohammadpur, a slum settlement in Dhaka, are working with a local NGO, Community Participation and Development, to trial the traditional community development and DRR tools to see which participatory methods and exercises are applicable to their urban context, which need to be revised and, consequently, whether entirely new methods need to be developed. Outcomes to date indicate that some tools, like community mapping to highlight hazard prone areas and safe spaces, are directly transferable. Other tools, like the seasonal calendar, have proved surprisingly useful to

highlight changes in rainfall patterns and associated flooding. But traditional consultation techniques (relying on getting groups of people to gather in a central location) need to be adapted for urban communities, which are generally more populous, less structured and more time constrained (Save the Children, 2013d).

Children's playfulness and creativity can also lead to the development of new methods or new approaches to assessing vulnerability and capacity in a changing climate. In India, children are using Lego blocks to create three-dimensional models of their urban communities, using different colours to identify risky and safe areas. The advantage of this technique is that the map can be revisited periodically to ensure it remains an accurate reflection of the community structure. The model also facilitates children's ability to communicate their views on risks and resources and to physically explore the impacts of disasters on their community. Involving children themselves in testing the applicability of common methodologies for their unique situation has resulted in a more context-appropriate and targeted approach to knowledge building and action planning than may otherwise have been the case.

Harnessing children's creativity to foster innovations in adaptation

Children are often willing and able to use materials at hand in innovative ways to increase their resilience and help their communities. In Kenya, a climate change project that worked directly with students in 40 schools to promote understanding of climate change impacts and catalyse micro-scale actions to increase resilience in school communities, resulted in near zero-cost innovations that have reverberated into the wider communities. On learning about the impacts of climate change on water and food security, children in a number of schools began using "gunny sacks" (large bags used to transport grain) to grow vegetables. The sacks require less space than a traditional garden and also use less water. Parents and community members began adopting the practice as the gunny sacks fit well in small courtyard spaces. Children also began developing larger school kitchen gardens in which they experimented with drought tolerant crops. In at least one school (Boy's Town school in Garissa) community members, many of whom are transitioning pastoralists, engaged with the children to learn how to grow a wider variety of crops. Community members also started purchasing surplus vegetables from the school garden (InterClimate Network, 2012).

Children also began collecting charcoal dust, waste paper and other combustible waste and making briquettes for cook stoves. The new briquettes burn cleaner than charcoal or cow dung and help reduce deforestation. A graduating student has turned the briquette-making into a microenterprise, selling briquettes in the local market.

Several schools started collecting used plastic bags and weaving them into attractive and durable shopping bags. Others cut plastic bottles to shape and used them as improvised guttering and downpipes on school buildings to increase water catchment. The additional water was, in some places, used to irrigate school kitchen gardens. These low and often no cost initiatives have started to generate change in the schools and surrounding communities, with families and communities benefiting from and often adopting the children's ideas (InterClimate Network, 2012). With climate change projected to increase climate variability across Kenya and reduce rainfall in the already arid and semi-arid parts of the country (Government of Kenya, 2013; World Bank, 2009), innovative, low-cost solutions to increasing food security and reducing the use of resources can play a key role in increasing local level resilience.

Sometimes disasters can motivate innovation. In Vietnam, in 2006, 19 children in Nghe An Province died when the boat they were taking to school capsized after heavy rains caused the river to rise rapidly. This is not an unusual occurrence in the Mekong Delta, but when these deaths were reported on local television an 11-year-old boy named Hieu, decided something had to be done. Hieu was inspired to create change by the fact that many of the children were still wearing their school backpacks when they were pulled from the river. The idea of the floating backpack was born. Hieu worked to design a prototype, which went through several iterations to find the right combination of safety, buoyancy and utility as a bag in a look that would not be shunned by school children. Once the design was finalized, Hieu convinced his mother to start a business manufacturing and distributing the backpacks. The national government endorsed the backpacks and a number of schools in flood prone areas have provided them to students. The backpack won a national design contest and an international design prize. Several private and civil society organizations have distributed the backpacks to children in flood prone regions, increasing safety (UNDP Vietnam, 2012). Climate change is projected to increase the frequency and severity of extreme rainfall events in Vietnam with consequent increases in flooding and increased river flows (Institute of Strategy and Policy on Natural Resources and Environment, 2009). This will result in more scenarios like the one that resulted in the deaths of those 19 children. While large scale adaptation measures are likely to be required in Vietnam's Mekong Delta provinces to increase flood protection, low-cost innovations like floating backpacks have a significant role to play, particularly in remote regions underserved by infrastructure.

Conclusions

Analysis of the case studies shows that in many instances, children are highly capable of developing innovative, low-cost solutions to real-world challenges. They can quickly grasp complex concepts and develop action-oriented strategies to reduce risk and take advantage of opportunities. But children cannot create and sustain the required change by themselves, and children's action groups for climate change need to be embedded into existing structures to ensure that knowledge and activities do not get lost when children graduate or when individual projects conclude (Tanner & Seballos, 2012). Approaches to building children's adaptive capacity should be institutionalized to ensure sustainability, and all action in this area should be child-centred to ensure all the insights, energy and knowledge children bring to this issue are captured, and to ensure that children's rights (as children and future adults) are protected in a changing climate. Given these findings it seems clear that mainstreaming climate risk and adaptation into sector- or issue-based projects working with children at the community level is likely to be a more effective means of ensuring sustainable adaptation outcomes than establishing stand-alone adaptation projects. However, further evidence needs to be gathered to support this conclusion.

The key outcome of this brief analysis is that much more research is required in the area of child-centred approaches to climate change action. Children's experience of, and capacity to manage, climate change is very under-represented in the literature. However, as highlighted above, there is a growing body of anecdotal evidence that taking a child-centred approach to increasing understanding and action on climate change can not only build the adaptive capacity of the children involved, but also provide benefits to entire communities. Yet there is no solid evidence-base proving that what has worked in a growing number of cases is more broadly applicable, translatable to other regions, or sustainable in the absence of direct project support. Collaborative efforts between researchers and practitioners could provide this evidence.

Note

1. This definition is recognized by every UN member state except Somalia and the USA which have not ratified the Convention.

References

Akachi, Y., Goodman, D., & Parker, D. (2009). *Global climate change and child health: A review of pathways, impacts and measures to improve the evidence base* (Innocenti Discussion Paper No. IDP 2009-03). Florence: UNICEF Innocenti Research Centre.

Anderson, A. (2010). *Combating climate change through quality education*. Washington, DC: The Brookings Institution.

Bangay, C., & Blum, N. (2010). Education responses to climate change and quality: Two parts of the same agenda? *International Journal of Educational Development, 30*(4), 335–450.

Bartlett, S. (2008). Climate change and urban children: Impacts and implications for adaptation in low- and middle-income countries. *Environment and Urbanization, 20,* 501–519.

Costello, A., Abbas, M., Allen, A., Ball, S., Bell, S., Bellamy, R., ... Patterson, C. (2009). Managing the health effects of climate change. *The Lancet, 373,* 1693–1733.

Dalal-Clayton, B., & Bass, S. (2009). *The challenge of environmental mainstreaming.* London: International Institute for Environment and Development.

Doherty, T., & Clayton, S. (2011). The psychological impacts of global climate change. *American Psychologist, 66*(4), 265–276.

Farrant, B., Armstrong, F., & Albrecht, G. (2012). Future under threat: Climate change and children's health. *The Conversation.* Retrieved June 20, 2013, from http://theconversation.com/future-under-threat-climate-change-and-childrens-health-9750

Foundation of the Peoples of the South Pacific International, & Plan International. (2013, June). *Children and climate change.* Newsletter Issue No. 2.

Frumkin, H., Hess, J., Luber, G., Malilay, J., & McGeehin, M. (2008). Climate change: The public health response. *American Journal of Public Health, 98*(30), 435–445.

Gaillard, J.C., & Pangilinan, M. (2010). Participatory mapping for raising disaster risk awareness among the youth. *Journal of Contingencies and Crisis Management, 18*(3), 175–179.

Government of Kenya. (2013). *National climate change action plan 2013–2017.* Nairobi: Author.

Huq, S., & Reid, H. (2004). Mainstreaming adaptation in development. *IDS Bulletin, 35*(3), 15–21.

Institute of Development Studies. (2009). Children communicating climate and disaster risks. *IDS in Focus Policy Briefing,* (13). Retrieved from https://www.ids.ac.uk/files/dmfile/IF13.3.pdf

Institute of Strategy and Policy on Natural Resources and Environment. 2009. *Viet Nam assessment report on climate change.* Hanoi: Author.

InterClimate Network. (2012). *Kenya impact report 2008–2011: International climate challenge.* Retrieved August 12, 2013, from http://www.blurb.co.uk/books/3228343-icc-impact-report-kenya-2008-2011?redirect=true

Intergovernmental Panel on Climate Change – Adger, W.N., Agrawala, S., Mirza, M.M.Q., Conde, C., O'Brien, K., Pulhin, J., ... Takahashi, K. (2007a). Assessment of adaptation practices, options, constraints and capacity. In M.L. Parry, O.F. Canziani, J.P. Palutikof, P.J. van der Linden, & C.E. Hanson (Eds.). *Climate Change 2007: Impacts, adaptation and vulnerability. Contribution of Working Group II to the Fourth Assessment Report of the Intergovernmental Panel on Climate Change* (pp. 719–743). Cambridge: Cambridge University Press.

Intergovernmental Panel on Climate Change – Confalonieri, U., Menne, B., Akhtar, R., Ebi, K.L., Hauengue, M., Kovats, R. S., ... Woodward, A. (2007b). Human health. In M.L. Parry, O.F. Canziani, J.P. Palutikof, P.J. van der Linden, & C.E. Hanson (Eds.). *Climate Change 2007: Impacts, adaptation and vulnerability. Contribution of Working Group II to the Fourth Assessment Report of the Intergovernmental Panel on Climate Change* (pp. 391–431). Cambridge: Cambridge University Press.

Intergovernmental Panel on Climate Change – Pachauri, R., & Reisinger, A. (Eds.). (2007c). *Contribution of Working Groups I, II and III to the Fourth Assessment Report of the Intergovernmental Panel on Climate Change.* Geneva: IPCC.

Intergovernmental Panel on Climate Change – Field, C.B., Barros, V., Stocker, T.F., Qin, D., Dokken, D.J., Ebi, K.L., ... Midgley, P.M. (Eds.). (2012). *Managing the risks of extreme events and disasters to advance climate change adaptation: A special report of Working Groups I and II of the Intergovernmental Panel on Climate Change.* Cambridge: Cambridge University Press.

Klein, R., Schipper, E.L.F., & Dessai, S. (2003). *Integrating mitigation and adaptation into climate and development policy: Three research questions* (Tyndall Centre Working Paper No. 40). Tyndall Centre for Climate Change Research.

Lansdown, G. (2001). *Promoting children's participation in democratic decision-making.* Florence: UNICEF.

Moser, C., & Moser, A. (2005). Gender mainstreaming since Beijing: A review of success and limitations in international institutions. *Gender and Development, 13*(2), 11–22.

Plan International. (2011). *Weathering the storm: Adolescent girls and climate change.* Woking: Author.

Plush, T. (2009). Amplifying children's voices on climate change: The role of participatory video. *Participatory Learning and Action, 60,* 119–128.

Save the Children. (2009). *Feeling the heat: Child survival in a changing climate.* London: International Save the Children Alliance.

Save the Children. (2012). Internal project reporting for *DRR Education* project.

Save the Children. (2013a). *Reducing risk, enhancing resilience.* London: International Save the Children Alliance.

Save the Children. (2013b). Internal project reporting for the *Child Centered – Community Based Climate Change Adaptation* project.

Save the Children. (2013c). Internal project reporting for *"Yumi stap redi long Climate change" The Vanuatu NGO Climate Change Adaptation Program.*

Save the Children. (2013d). Internal project reporting for *Integrated Child Centred Climate Change Adaptation in Bangladesh* project.

Schipper, E.L.F. (2007). *Climate change adaptation and development: Exploring the linkages* (Tyndall Centre Working Paper No. 107). Tyndall Centre for Climate Change Research.

Seballos, F., Tanner, T., Tarazona, M., & Gallegos, J. (2011). *Children and disasters: Understanding impact and enabling agency.* Brighton: Institute of Development Studies.

Sheffield, P., & Landrigan, P. (2011). Global climate change and children's health: Threats and strategies for prevention. *Environmental Health Perspectives, 119*(3), 291–298.

Smit, B., & Wandel, J. (2006). Adaptation, adaptive capacity and vulnerability. *Global Environmental Change, 16*(2006), 282–292.

Tanner, T. (2010). Shifting the narrative: Child-led responses to climate change and disasters in El Salvador and the Philippines. *Children & Society, 24*(4), 339–351.

Tanner, T., Garcia, M., Lazcano, J., Molina, F., Molina, G., Rodriguez, G., ... Seballos, F. (2009). Children's participation in community-based disaster risk reduction and adaptation to climate change. *Participatory Learning and Action, 60,* 54–64.

Tanner, T., & Seballos, F. (2012). Children, climate change and disasters. *IDS in Focus: Policy Briefing,* (23). Retrieved from https://www.ids.ac.uk/files/dmfile/IF23.pdf

Telford, J., Cosgrave, J., & Houghton, R. (2006). *Joint evaluation of the international response to the Indian Ocean tsunami: Synthesis report.* London: Tsunami Evaluation Coalition.

Turnbull, M., Sterrett, C., & Hilleboe, A. (2013). *Towards resilience: A guide to disaster risk reduction and climate change adaptation.* Rugby: Practical Action.

UNEARTH News. (2013). *Climate change education empowers children in Vietnam*. Retrieved August 5, 2013, from http://unearthnews.org/climate-change-education-empowers-children-in-vietnam/

United Nations. (1989). *Convention on the Rights of the Child*. Retrieved June 20, 2013, from http://www.ohchr.org/EN/ProfessionalInterest/Pages/CRC.aspx

UN. (1992). *United Nations Framework Convention on Climate change*. Retrieved September 10, 2013, from http://unfccc.int/files/essential_background/background_publications_html pdf/application/pdf/conveng.pdf

United Nations Children's Fund. (2007). *Climate change and children*. New York: Author.

United Nations Children's Fund. (2011). *Children's vulnerability to climate change and disaster impacts in East Asia and the Pacific*. Bangkok: UNICEF East Asia and Pacific Regional Office.

United Nations Children's Fund, & Plan International. (n.d.). *The benefits of a child-centred approach to climate change adaptation*. Retrieved June 20, 2013, from http://www.childreninachangingclimate.org/database/plan/Publications/The-Benefits-of-a-child-centred-appraoch-to-climate-change-adaption.pdf

United Nations Development Programme & United Nations Environment Programme. (2011). *Mainstreaming climate change adaptation into development planning: A guide for practitioners*. Retrieved December 10, 2013, from http://www.unpei.org/pei-pep-publications

United Nations Development Programme Vietnam. (2012). *Floating backpack helps children get back to school*. Retrieved August 12, 2013, from http://www.undp.org/content/vietnam/en/home/ourwork/environmentclimate/successstories/successstories/

United Nations Office for Disaster Risk Reduction, & Plan International. (2012). *Children's action for disaster risk reduction: Views from children in Asia*. Bangkok: UNISDR and Plan Asia Regional Office.

Waterson, T. (2006). Climate change – the greatest crisis for children? *Journal of Tropical Pediatrics, 52*(6), 383–385.

World Bank. (2009). *Making development climate resilient: A World Bank strategy for sub-Saharan Africa*. Washington, DC: Author.

CASE STUDY

Knowledge flows in climate change adaptation: exploring friction between scales

Clare Stott[a,b] and Saleemul Huq[b,c]

[a]Department of Anthropology, University College London, London, UK; [b]International Centre for Climate Change and Development, Dhaka, Bangladesh; [c]International Institute for Environment and Development, Climate Change Group, London, UK

abstract>
Effective mainstreaming of climate change adaptation (CCA) into related policy and development initiatives relies on comprehensive knowledge sharing between multiple stakeholders. In Bangladesh, community-based adaptation (CBA) practitioners are critical for facilitating communication among global, national and local scales. They can also take responsibility for finding appropriate channels through which to share relevant information. Interviews with CBA practitioners examine how knowledge is gained and transmitted between practitioners and other CCA stakeholders, focusing on the challenges experienced. These challenges represent friction in knowledge transmittal. Key to lubricating smooth knowledge flows is an understanding of the specific contexts within which knowledge is to be exchanged. At the professional level, multidisciplinary knowledge must be made accessible through provision of widely comprehensible content shared in an appropriate format. At the local level, understandings of trust, priorities and power relations are vital for ensuring relevance in the knowledge shared by professional stakeholders. Mobilizing appropriate knowledge can allow widespread comprehension of adaptation aims, enabling the mainstreaming of CCA and ensuring that resulting action is beneficial at the local level, for communities that are vulnerable to the impacts of climate change.

1. Introduction

Climate change is an international problem discovered by the global scientific community. With widely agreed certainty about its damaging effects, efforts are being made to increase climate change adaptation (CCA) action. This process comprises adjustment in human and natural systems in order to cope with or benefit from environmental changes (Smit et al., 1999). It is acknowledged that local communities develop their own coping strategies, yet there is a notion that specialist support must be provided due to the pace of witnessed and predicted climatic change. As such, anthropogenic adaptation is receiving much consideration from climate change scientists (Grothmann & Patt, 2005), while global climate policy aims to bring adaptation to the centre of the contemporary climate change discourse (Adger et al., 2009).

This focus enables mainstreaming of CCA into policy and development initiatives. The problem of climate change is brought to the attention of policy-makers through the reports of the Intergovernmental Panel on Climate Change. Associated global policy-making takes place under the aegis of the United Nations Framework Convention on Climate Change (Dirix et al., 2013). Such policy affects the extent of national and international support for CCA and encourages the prioritization of adaptation activities within governance and development approaches.

In order to be effective, CCA mainstreaming relies upon widespread comprehension of adaptation aims and issues at multiple scales. This requires strong communicative connections between multi-scalar CCA stakeholders (Crate, 2011). Stakeholders include climate scientists, national and international policy-makers, government bodies, research institutions, development workers, community groups, and, critically, those who are vulnerable to the impacts of climate change and, as such, need to adapt. Knowledge sharing between stakeholders can create and maintain connections, inform effective policy and ensure that emerging policies benefit those communities who are under pressure to adapt. However, integration between multiple scales is a complex ambition characterized by friction (Tsing, 2005). To address this, reciprocal communications along 'strong chains' with 'short links' (Brown & Fox, 1999) are beneficial (Agar,

boilerplate>
This is an Open Access article. Non-commercial re-use, distribution, and reproduction in any medium, provided the original work is properly attributed, cited, and is not altered, transformed, or built upon in any way, is permitted. The moral rights of the named author(s) have been asserted.

92

2005; Blackburn & Clark, 2007; Peacock, 2010; Root & Schneider, 1995). They can encourage smooth knowledge flows between stakeholders, achieving widespread comprehension and, thereby, supporting effective mainstreaming of CCA.

While some channels of communication are well established, such as those from global scientists to global policy-makers, others are not. The weakest link is with vulnerable communities at the local level, in terms of both gaining their knowledge of local climate and adaptation and sharing climate change knowledge with them. The global focus on CCA manifests itself at the local level through community-based adaptation (CBA) projects. CBA projects are designed to promote and protect sustainable livelihoods through building capacity for community-level adaptation to unpredictable and risky climates (Ayers & Forsyth, 2009; Ireland & McKinnon, 2013). Practitioners working within this field provide a direct link with local communities. As such, they can operate as vehicles for the reciprocal transmittal of knowledge between scales and stakeholders. This case study explores knowledge sharing in relation to CCA from a national-level practitioner perspective in Bangladesh, drawing from research conducted with CBA practitioners from the national platform 'Action Research for Community Adaptation in Bangladesh' (ARCAB).

2. Knowledge exchange in Bangladesh

Bangladesh provides a national platform upon which to explore CBA. Recognition of the country's vulnerability to the impacts of climate change means that much internal adaptation research and implementation is occurring (Rashid, Khan, & Khan, 2013). The lessons learnt through this work can be applied in similarly vulnerable countries. Within Bangladesh, there is a wide network of developmental organizations (Zohir, 2004). Strong partnerships among these organizations can provide a pivotal portal for the crucial exchange of ideas, strategies and viewpoints (Benson, Twigg, & Myers, 2001).

ARCAB is a programme directed by the International Centre for Climate Change and Development in Dhaka, which brings together national and international organizations to research and implement local-level CBA activities. In doing so, it provides a platform upon which CBA practitioners exchange knowledge. As ARCAB organizations are incorporating CBA initiatives into their central development activities, associated practitioners can take responsibility for gaining and transmitting appropriate climate change knowledge among the vulnerable communities within which they work. In turn, they can also feed back to political stakeholders at national and international scales to advocate for adaptation needs and measures for consideration in policies. There is growing appreciation within these organizations that knowledge management

must be based on inclusive access to trustworthy, effective and widespread communications (Action Aid Bangladesh, 2012; Practical Action, 2013). However, problems remain due to complex contrasts between the specific contexts within which knowledge for CBA is generated and shared.

The discussion that follows is based on qualitative research conducted with 12 CBA practitioners from the national offices of 12 international ARCAB partners. One-to-one semi-structured interviews were employed to critically examine the challenges experienced in gaining CCA knowledge and sharing it at the local level. These participants were strategically targeted with the knowledge that ARCAB institutions are directly involved with CBA and knowledge sharing around CBA. It is acknowledged that in sampling solely from ARCAB partners, a potential diversity of communication methods employed by different institutions involved in CBA in Bangladesh may not be represented and, as such, the results cannot claim to be wholly representative of the national-level practitioner community. Nevertheless, these interviews have revealed some significant frictions that hinder smooth knowledge flows between multiple scales, in the context of CBA (see Figure 1).

3. Results

3.1. *Practitioner acquisition of climate and adaptation information*

While ARCAB practitioners reported no problems with gaining knowledge from local communities, they identified some challenges with regard to gaining climate change and adaptation knowledge from alternative sources. Information to support local adaptation strategies is sought from a variety of sources. These include national government ministries, national and international knowledge platforms, national and international non-governmental organizations (NGOs), universities, scientific bodies, research institutions, media and internal research. The challenges identified are indicated in Figure 1 and are discussed further below to illuminate how they disrupt smooth knowledge flows.

3.1.1. *Personal networks*

The personal networks of NGO practitioners were identified as hugely significant in gaining climate change knowledge and information. Personal networks interlink practitioners with multi-scale knowledge sources, including partner organizations, other NGOs, research institutes, government members and local-level stakeholders. The professional relationships developed throughout individual practitioners' careers are pivotal to the acquisition of knowledge and support. As such, when employees leave their role, the connection between the organization and

↓ Friction ↓   ↓ Friction ↓

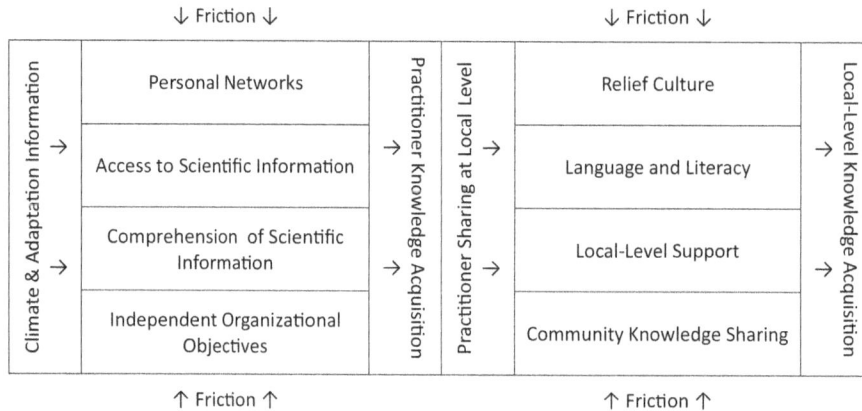

↑ Friction ↑   ↑ Friction ↑

Figure 1.   Frictions affecting smooth knowledge flows to and from CBA practitioners.

knowledge source is disrupted. Consequently, the institution's ability to gather essential information is weakened. Longer serving employees who have developed extensive networks are likely to have more leverage within their role (Benson et al., 2001). This leverage can increase efficiency in gaining appropriate climate change knowledge. It is inevitable that the personal circumstances of CBA practitioners will change depending on individual desires, opportunities and situations. Hence, it is crucial that development organizations explore practical avenues for the transfer of information and contacts across programmes and through successive generations of staff (Benson et al., 2001).

3.1.2.   *Access to scientific information*

Practitioners reported access limitations as an obstacle to gaining scientific climate change information. While models of specific impacts may not be necessary to inform policy (Dessai & Hulme, 2004) there is pressure on practitioners to provide relevant and localized information to communities. The premise of this is that individuals are more likely to adapt if they have some certain, tangible and reliable information about probable climatic impacts on their livelihoods (Adger et al., 2009). Practitioners reported that where such knowledge exists, it is publicly unavailable and, hence, unobtainable for these development organizations. Such shortcomings reflect an acknowledged institutional disconnect between scientific bodies and local-level stakeholders (Benson et al., 2001).

3.1.3.   *Comprehension of scientific information*

Some practitioners reported that their scientific understanding of climate change is not as comprehensive as they desire. Within CBA there is an overriding focus on social issues relating to adaptation. However, a restricted understanding of the scientific forces at play in climate change

can hinder a full comprehension and, in turn, can weaken the ability to explain related concepts to a wider audience. As such, the knowledge of climate change scientists is not easily transferable to practitioners. These limitations cause frustration and, in the long term, a lack of motivation to find the specific information that could help to inform adaptation programmes.

3.1.4.   *Independent organizational objectives*

Interviewees reported difficulties in gaining knowledge from other development organizations. ARCAB partners rarely consider CCA in isolation. Instead, they aim to mainstream CCA within existing development agendas. Each organization has its own commitments, defined by certain aims and principles and the expectations of the donors that are financing their programmes (Stirrat & Henkel, 1997). This self-guided aspect to each organization can detract from using and sharing appropriate knowledge within and between collaborative knowledge platforms.

3.2.   *Sharing knowledge at the local level*

Within ARCAB there is no standard model for sharing climate change knowledge with local communities. Practitioners reported different channels and techniques, each with their own successes and challenges. Face-to-face communication is the main means of both sharing and gaining knowledge at the local level. These personal interactions occur between local partner NGO field staff, community volunteers, children, community groups and government extension officers. Personal interaction is supported through various media, employing visual (flip charts, posters, maps), written (documents, school books), audio (loudspeaker) and interactive (knowledge centres, drama) channels. Yet, face-to-face communication is considered the most effective, taking precedence over communication through written material (Benson et al., 2001). The frictions apparent to disrupting smooth knowledge flows between CBA practitioners and local-level stakeholders are outlined in Figure 1 and explored below.

3.2.1. *Relief culture*

Practitioners reported that a significant obstacle to adequately sharing knowledge at the local level is an expectation and desire for relief within vulnerable communities. This expectation overcomes the desire to learn skills and develop the theoretical knowledge offered by development organizations. Instead, it was reported that community members are more willing to act if they expect to receive something tangible in return. In acknowledgement of this, participation in adaptation activities is encouraged by small material provisions. However, persistent dependence on official aid can ultimately disempower communities (Rowlands, 1995; Toomey, 2011). The focus remains on the aid being received rather than the knowledge being shared and can therefore negatively impact the sustainability of adaptation activities. As such, this 'relief culture' poses a major obstacle for developmental organizations working on adaptation (Benson et al., 2001).

3.2.2. *Language and literacy*

It is obvious that language is a pivotal aspect in any type of knowledge exchange, yet several language issues were identified. First, it was reported that the technical terminology that is used in climate change discussions is hard to translate. The inclusion of technical terms and external concepts is often uncomfortable and unappealing to stakeholders at the community level. Second, participants suggested that a balance must be maintained with regard to the complexity of information shared. Their experience found that information that is too simplistic and broad is as ineffective as information that is too complex and abstract to community perspectives. Third, ARCAB practitioners gain much internationally generated information in English. Some organizations have shared knowledge with local communities without translating it into the appropriate dialect. Literacy and language barriers often exclude certain groups within societies, such as the rural poor, ethnic minorities and women (Roncoli, 2006). This goes against the principles of adaptation, which should target the most vulnerable (Rashid et al., 2013). Hence, rather than aiding the adaptation process, poorly translated information can serve as a deterrent, alienating people who make no connection with the material. These issues indicate that it is vital to account for the specific linguistic context within which knowledge is being shared.

3.2.3. *Local-level support*

Local-level knowledge transmission from ARCAB partners is supported by local-level stakeholders. For example, CBA practitioners work alongside local government officials and field extension staff to provide agricultural adaptation advice. However, each actor assuming some responsibility for transmitting knowledge must be working towards similar aims in order to avoid confusion and ambiguity. This research found that government information is sometimes outdated. While CBA practitioners can work with governments to ensure agreement in the knowledge being employed, governments must also work to ensure that they access and make available the most up-to-date information. This requires the whole process of knowledge sharing to include a feedback mechanism between actors at the local level and beyond in order to ensure mutual understanding between stakeholders.

Field staff members operating at the local level are often employed from within vulnerable communities in acknowledgment that knowledge must be transmitted via reliable and trustworthy sources. The familiarity between such field staff and individuals within target communities is likely to aid the adaptation process and boost trust in the climate knowledge that is shared (Ensor & Berger, 2009). Yet the interviews indicated that here, again, the turnaround of staff can disrupt networks of personal interaction. This can affect levels of trust, which are pivotal to comprehensive knowledge sharing processes (Hasnain & Jasimuddin, 2012).

3.2.4. *Community knowledge sharing*

Knowledge sharing within the community itself was indicated to be a further challenge. It is expected that once useful information has been identified and acquired, community members will share it with neighbours, friends and acquaintances who are experiencing similar problems. Hence, development practitioners perceive their direct audiences to be accountable for mobilizing widespread knowledge flows at the local level. However, some participants reported that this is not always a reliable assumption, explaining that knowledge is not passed on beyond the initial recipients. Information deemed to be irrelevant or abstract by the community may not be shared among its members (Adhikari & Taylor, 2012; Agar, 2005). Instead, individuals are likely to share information that is genuinely deemed to be interesting, relevant and useful to their livelihoods. Therefore, knowledge flows can be enhanced through ensuring contextual relevance in the information shared.

4. Ensuring smooth knowledge flows

Ensuring smooth knowledge flows to enable and enhance the mainstreaming of CCA into policy and development is a challenging ambition. It requires flexibility, commitment and persistence from a variety of stakeholders that operate at different scales. These multi-scalar collaborations encompass a disparity in perspectives about what adaptation should be and how it should be prioritized and executed (Agar, 2005; Nuttall, 2009). Moreover, as these interviews have indicated, friction in gaining appropriate knowledge remains apparent, at least for adaptation practitioners. Such frictions continually disrupt and destabilize knowledge flows (Tsing, 2005). However, smooth knowledge flows can be encouraged by accounting for the particular social, political and economic context within which the recipient exists (Action Aid Bangladesh, 2012; Adhikari & Taylor, 2012; Magistro & Roncoli, 2001) to ensure that both the format and content of knowledge shared is appropriate.

A comprehension of specific contexts is particularly crucial for local-level knowledge sharing. In employing face-to-face communication techniques there must be a level of trust between recipient and knowledge provider (Vignola et al., 2013). Social research has highlighted a lack of local-level trust in global information (Ellis, Cambray, & Lemma, 2013). Moreover, at the local level, such global knowledge does not remain neutral, but has been found to insert itself into existing power relations (Broad & Orlove, 2007; Crewe & Young, 2002; Dankelman, 2002; Denton, 2002; Peacock, 2010; Roncoli, 2006). Hence, instead of precise external scientific knowledge, specific worldviews, robust decision-making and personal motivations may be more important to achieving sustainable adaptation (Adger et al., 2009; Grothmann & Patt, 1995; Nuttall, 2009). It becomes apparent that the social and political dynamics of communities within which knowledge is being shared is as significant as the ecological and the economic contexts for ensuring relevance of content and widespread comprehension of adaptation aims.

The disparity between perspectives at different scales poses the question: how can we elicit sympathy for short-term perspectives within the international climate change community, while also encouraging long-term outlooks among local-level stakeholders? Many have suggested that multi-sited ethnography can be employed to understand critical global–local linkages (Crate, 2011; Henning, 2005; Krauss, 2009; Marcus, 1995; Nelson, West, & Finan, 2009; Smart, 1999). Such an approach can serve to encompass scientific, practical and philosophical perspectives related to the various scenarios of climate change. Future research in Bangladesh and other climatically vulnerable countries can focus on ensuring contextual comprehension of specific localities.

By enhancing the flow of relevant knowledge, mainstreaming of CCA can be more effective. Many countries implementing CBA initiatives lack the crucial networks of stakeholders necessary to connect the gaps between global and local. Globally, the responsibilities of specific government and non-government actors have been blurred and fragmented, leaving ambiguity as to who is accountable for ensuring cohesion between scales (Dirix et al., 2013). Communicative connections need to be addressed at each link to achieve appropriate and impactful knowledge sharing and, in turn, ensure smooth and effective knowledge flows. Contributors to adaptation knowledge from a range of disciplinary backgrounds can take responsibility for making appropriate knowledge accessible. Recipients can thereby combine scientific, political and social sciences to inform adaptation policy, projects and knowledge (Sutton, 1999). CBA practitioners, local government officials and other adaptation stakeholders interacting at the local level can ensure that adequate information is collated and integrated into effective responses via policy and action, in order to support adaptation in various contexts. Success in this can ensure that global policies respond to local priorities and, thereby, that the political support achieved through mainstreaming adaptation is effective at the local level.

Acknowledgements

Support and guidance for this project came from colleagues and contacts at the International Centre for Climate Change and Development in Dhaka and the Institute of Environment and Development in London. Gratitude is extended to the participants of this research.

References

Action Aid Bangladesh. (2012). *Scaling up community based adaptation with local government.* Project Completion Report. Dhaka: ActionAid Bangladesh.

Adger, W.N., Dessai, S., Goulden, M., Hulme, M., Lorenzoni, I., Nelson, D.R., … Wreford, A. (2009). Are there social limits to adaptation to climate change? *Climatic Change, 93*(3–4), 335–354. doi:10.1007/s10584-008-9520-z

Adhikari, B., & Taylor, K. (2012). Vulnerability and adaptation to climate change: A review of local actions and national policy response. *Climate and Development, 4*(1), 54–65. doi:10.1080/17565529.2012.664958

Agar, M. (2005). Local discourse and global research: The role of local knowledge. *Language in Society, 34*, 1–22. doi:10.1017/S0047404505050013

Ayers, J., & Forsyth, T. (2009). Community-based adaptation to climate change: Strengthening resilience through development. *Environment: Science and Policy for Sustainable Development, 51*(4), 22–31.

Benson, C., Twigg, J., & Myers, M. (2001). NGO initiatives in risk reduction: An overview. *Disasters, 25*(3), 199–215. doi:10.1111/1467-7717.00172

Blackburn, M.V., & Clark, C.T. (2007). Bridging the local/global divide: Theorizing connections between global issues and local actions. In M.V. Blackburn & C.T. Clark (Eds.), *Literacy research for political action and social change* (pp. 9–28). New York, NY: Peter Lang.

Broad, K., & Orlove, B. (2007). Channeling globality: The 1997–98 El Niño climate event in Peru. *American Ethnologist, 34* (2), 285–302. doi:10.1525/ae.2007.34.2.285

Brown, L., & Fox, J. (1999). *Transnational civil society coalitions and the World Bank: Lessons from project and policy influence campaigns.* Rochester, NY: Social Science Research Network.

Crate, S.A. (2011). Climate and culture: Anthropology in the era of contemporary climate change. *Annual Review of Anthropology, 40*(1), 175–194. doi:10.1146/annurev.anthro.012809.104925

Crewe, E., & Young, J. (2002). *Bridging research and policy: Context, evidence and links* (Overseas Development Institute, Working Paper 173).

Dankelman, I. (2002). Climate change: Learning from gender analysis and women's experiences of organising for sustainable development. *Gender and Development, 10*(2), 21–29. doi:10.1080/13552070215899

Denton, F. (2002). Climate change vulnerability, impacts, and adaptation: Why does gender matter? *Gender & Development, 10*(2), 10–20. doi:10.1080/13552070215903

Dessai, S., & Hulme, M. (2004). Does climate adaptation policy need probabilities? *Climate Policy, 4*(2), 107–128. doi:10.1080/14693062.2004.9685515

Dirix, J., Peeters, W., Eyckmans, J., Jones, P.T., & Sterckx, S. (2013). Strengthening bottom-up and top-down climate governance. *Climate Policy, 13*(3), 363–383. doi:10.1080/14693062.2013.752664

Ellis, K., Cambray, A., & Lemma, A. (2013). *Drivers and challenges for climate compatible development.* London: Climate and Development Knowledge Network.

Ensor, J., & Berger, R. (2009). *Understanding climate change adaptation: Lessons from community-based approaches.* Rugby: Practical Action.

Grothmann, T., & Patt, A. (2005). Adaptive capacity and human cognition: The process of individual adaptation to climate change. *Global Environmental Change, 15*(3), 199–213. doi:10.1016/j.gloenvcha.2005.01.002

Hasnain, S.S., & Jasimuddin, S.M. (2012). Barriers to knowledge transfer: Empirical evidence from the NGO (non-governmental organizations)-sector in Bangladesh. *World Journal of Social Sciences, 2*(2), 135–150.

Henning, A. (2005). Climate change and energy use: The role for anthropological research. *Anthropology Today, 21*(3), 8–12. doi:10.1111/j.0268-540X.2005.00352.x

Ireland, P., & McKinnon, K. (2013). Strategic localism for an uncertain world: A postdevelopment approach to climate change adaptation. *Geoforum, 47*, 158–166. doi:10.1016/j.geoforum.2013.01.005

Krauss, W. (2009). Localizing climate change: A multi-sited approach. In M.-A. Falzon (Ed.), *Multi-sited ethnography. Theory, praxis and locality in contemporary research* (pp. 149–164). Surrey: Ashgate.

Magistro, J., & Roncoli, C. (2001). Anthropological perspectives and policy implications of climate change research. *Climate Research, 19*(2), 91–96. doi:10.3354/cr019091

Marcus, G.E. (1995). Ethnography in/of the world system: The emergence of multi-sited ethnography. *Annual Review of Anthropology, 24*(1), 95–117.

Nelson, D.R., West, C.T., & Finan, T.J. (2009). Introduction to "In focus: Global change and adaptation in local places". *American Anthropologist, 111*(3), 271–274. doi:10.1111/j.1548-1433.2009.01131.x

Nuttall, M. (2009). Living in a world of movement: Human resilience to environmental instability in Greenland. In S. A. Crate & M. Nuttall (Eds.), *Anthropology and climate change: From encounters to actions* (pp. 292–310). Walnut Creek, CA: Left Coast Press.

Peacock, J. (2010). *Grounded globalism: How the U.S. south embraces the world.* Athens: University Of Georgia Press.

Practical Action. (2013). *Information, communication, learning.* Retrieved February 28, 2013, from https://practicalaction.org/information-communication-learning-answers

Rashid, M.A.K.M., Khan, M.R., & Khan, M.R. (2013). Community based adaptation: Theory and practice. In R. Shaw, F. Mallick, & A. Islam (Eds.), *Climate change adaptation actions in Bangladesh: Disaster risk reduction* (Part IV, pp. 341–362). Japan: Springer.

Roncoli, C. (2006). Ethnographic and participatory approaches to research on farmers' responses to climate predictions. *Climate Research, 33*(1), 81–99. doi:10.3354/cr033081

Root, T.L., & Schneider, S.H. (1995). Ecology and climate: Research strategies and implications. *Science, 269*(5222), 334–341. doi:10.1126/science.269.5222.334

Rowlands, J. (1995). Empowerment examined. *Development in Practice, 5*(2), 101–107. doi:10.1080/0961452951000157074

Smart, A. (1999). Participating in the global: Transnational social networks and urban anthropology. *City and Society, 11*(1–2), 59–77. doi:10.1525/city.1999.11.1-2.59

Smit, B., Burton, I., Klein, R.J., & Street, R. (1999). The science of adaptation: A framework for assessment. *Mitigation and Adaptation Strategies for Global Change, 4*(3–4), 199–213. doi:10.1023/A:1009652531101

Stirrat, R.L., & Henkel, H. (1997). The development gift: The problem of reciprocity in the NGO world. *The Annals of the American Academy of Political and Social Science, 554*(1), 66–80.

Sutton, R. (1999). *The policy process: An overview* (Overseas Development Institute Working Paper 118).

Toomey, A.H. (2011). Empowerment and disempowerment in community development practice: Eight roles practitioners play. *Community Development Journal, 46*(2), 181–195. doi:10.1093/cdj/bsp060

Tsing, A.L. (2005). *Friction: An ethnography of global connection.* Princeton, NJ: Princeton University Press.

Vignola, R., Klinsky, S., Tam, J., & McDaniels, T. (2013). Public perception, knowledge and policy support for mitigation and adaption to climate change in Costa Rica: Comparisons with North American and European studies. *Mitigation and Adaptation Strategies for Global Change, 18*(3), 303–323. doi:10.1007/s11027-012-9364-8

Zohir, S. (2004). NGO sector in Bangladesh: An overview. *Economic and Political Weekly, 39*(36), 4109–4113.

REVIEW ARTICLE

Up-scaling finance for community-based adaptation

Adrian Fenton[a,b,c], Daniel Gallagher[d], Helena Wright[c,e], Saleemul Huq[c,f] and Charles Nyandiga[g]

[a]Sustainability Research Institute, University of Leeds, Leeds, UK; [b]Centre for Climate Change Economics and Policy, Leeds, UK; [c]International Centre for Climate Change and Development, Dhaka, Bangladesh; [d]Adaptation Fund Board secretariat, Washington, DC, USA; [e]Imperial College London, Centre for Environmental Policy, London, UK; [f]International Institute for Environment and Development, London, UK; [g]United Nations Development Programme, NY, USA

While most adaptation actions occur at the local level, there is an absence of commitment at the international level to channel adaptation finance to local communities. Without such a commitment, there is a risk that climate finance will continue to support top-down, centralized activities that may struggle to address the needs of vulnerable communities. This paper explores ways in which community-based adaptation is presently being mainstreamed through the multilateral funds that are used to channel adaptation finance under the United Nations Framework Convention on Climate Change process, and points to two promising examples that demonstrate this. The first is the Small Grants Programme of the Global Environmental Facility, an established modality through which community organizations can access finance to manage their adaptation needs. The second is the direct access modality of the Adaptation Fund, which devolves decision-making power from multilateral agencies towards the national and local levels. At the country level, experiences from Nepal demonstrate an institutional environment that helps to prioritize the adaptation needs of the most vulnerable. Nepal achieves this by mandating that at least 80% of available finance flows to the community level, and that the implementation of projects is conducted in a bottom-up and inclusive process.

1. Introduction

The effects of rising temperatures, changing rainfall patterns, and extreme weather events exacerbated by climate change are causing severe and diverse impacts with implications for society and sustainable development (IPCC, 2012). Climate change poses a particular threat to developing countries, so developed countries have agreed under the United Nations Framework Convention on Climate Change (UNFCCC) to provide adaptation finance to help developing countries adapt to the impacts of climate change (UNFCCC, 2011).

In view of the increased adaptation finance expected to be raised by the international community under the UNFCCC, this paper explores ways in which adaptation finance has been channelled from finance institutions to community-based adaptation (CBA) initiatives to date, and the delivery mechanisms that have shown potential in up-scaling CBA.

Drawing on Schipper, Ayers, Reid, Huq, and Rahman (2014), we define up-scaling as horizontal replication and vertical mainstreaming of CBA efforts. Horizontal replication refers to the expansion or reproduction of existing efforts through multiple, small initiatives, while vertical mainstreaming refers to the integration of CBA into dominant policy and practice in local, regional, national, or even supra-national institutions (see Pelling, 2011). We recognize that many other definitions of up-scaling exist (see Schipper et al., 2014); however, for the purposes of this paper we focus on these elements.

Overall, we find that the majority of adaptation finance channelled through international mechanisms has been managed by multilateral entities without a strong focus on locally designed and locally implemented solutions that address the needs of the most vulnerable. This paper posits that significant resources must be channelled to developing countries not only at the national level, but also at the local level to enable adaptation to be effectively addressed.

2. CBA: top-down and bottom-up approaches

One of the ways in which adaptation efforts are characterized is either "bottom-up" or "top-down": a form of distinction whose origin lies in the policy implementation literature (Urwin & Jordan, 2008). For the purposes of this paper, top-down policy implementation refers to policies created by government, which tend to rely on expert technical advice, which are then operationalized at the local level. A bottom-up approach recognizes the importance of local context and other actors, particularly those at the level in which adaptation is operationalized, in formulating and implementing policies (see Urwin & Jordan, 2008; Van Aalst, Cannon, & Burton, 2008). In practice, the distinction between them can be blurred since bottom-up adaptation can be linked to government policies and plans and also follow set procedures.

CBA has often been referred to as a bottom-up approach (Sabates-Wheeler, Mitchell, & Ellis, 2008) and is becoming increasingly popular for operationalizing local inclusiveness (Ayers, 2011; Forsyth, 2013). CBA is different from top-down approaches as it is a community-led process, allowing climate vulnerability to be addressed at the local level in its specific context of impacts and adaptive capacity (Ayers & Forsyth, 2009; Forsyth, 2013). Building on the priorities, needs, knowledge, and capacities of communities, it empowers local-level climate-vulnerable stakeholders to be able to plan and cope with immediate climate variability and long-term climate change (Ensor & Berger, 2010; Reid et al., 2009). Climate-vulnerable stakeholders are groups or individuals vulnerable to the adverse effects of climate change. CBA centres on local knowledge of climatic hazards, as well as top-down model-based projections (Reid et al., 2009).

As climate change is only one of the stressors faced by climate-vulnerable stakeholders, interventions focusing exclusively on climate risks will be unlikely to reflect community priorities (Reid et al., 2009). CBA seeks to accommodate this by building upon participatory processes with local stakeholders, and development and disaster risk-reduction practitioners (Huq & Reid, 2007; Reid et al., 2009). Consequently, CBA often has additional poverty reduction and livelihood benefits, as well as reducing vulnerability to climate change and disasters (Reid et al., 2009). CBA may resemble typical development; however, the difference is that CBA incorporates in the potential impact of climate change on livelihoods and vulnerability to disasters (Reid et al., 2009).

It is unlikely that either form of adaptation will be sufficient in isolation from the other. For instance, justifications for financing top-down adaptation are often based on the constraints of bottom-up adaptation such as limited information, incentives, and access to resources which inhibit adaptation (Smit et al., 2001; Vernon, 2008). However, insufficient information held by governments on local issues also restricts possibilities for top-down adaptation, contributing to a need for bottom-up approaches (Moench & Dixit, 2004). Additionally, Moser and Satterthwaite (2008, p. 18) note that if

most city or municipal governments have proved unable or unwilling to provide the infrastructure, services, institutions and regulations to reduce risks from extreme weather events for much of their populations, they are unlikely to develop the capacity necessary to adapt to climate change.

Moreover, Adger, Huq, Brown, Conway, and Hulme (2003) noted that at times the state allows climate-related risks to exist as part of the politicized nature of urban planning and control. Consequently, it cannot be assumed that top-down approaches alone will reduce the vulnerability of marginalized communities. The deficiencies of top-down approaches have increased the relevance of bottom-up approaches at the local scale (Van Aalst et al., 2008). Much adaptation will inevitably continue to be bottom-up (Adger et al., 2003; Yamin, Rahman, & Huq, 2005) and these actions need support to ensure their effectiveness.

Currently, the dominant trend has been towards financing adaptation through top-down efforts (Schultz, 2012; Reid et al., 2009). This is perhaps because of the trend towards programme-based approaches requiring functioning national financial management systems along with rules, procedures, and transparent accounting (Brown & Kaur, 2009). Additionally, a lack of clear distinction between CBA and development (Ayers & Dodman, 2010) has perhaps contributed to the difficulty of financing CBA with official climate finance, which cannot be used to fund adaptation deficits (see Burton, 2004).

3. Overview of existing adaptation finance under the UNFCCC

At the time of writing, institutional arrangements for disbursing adaptation finance under the UNFCCC are in practice managed by the Global Environment Facility (GEF), the World Bank, and the Adaptation Fund Board. At COP-7 in 2001, the Marrakesh Accords (UNFCCC, 2002) declared the delivery of climate finance would be through replenishment of the GEF, bilateral and multilateral sources, and three specially set up funds: the Least Developed Country Fund (LDCF), the Special Climate Change Fund (SCCF), and the Kyoto Protocol Adaptation Fund (henceforth referred to as the Adaptation Fund). In addition to UNFCCC funds is the independent Pilot Programme for Climate Resilience (PPCR), a component of the Strategic Climate Fund (SCF), one of two funds within the World Bank's Climate Investment Funds (CIFs).

To date, the LDCF has allocated US$293.6 million for the preparation and implementation of National Adaptation Programmes of Actions (NAPAs), while the SCCF has allocated US$167.6 million for adaptation, technology transfer, and capacity-building (GEF, 2013a). The PPCR has allocated US$399 million for the integration of climate risk in national and sectoral development planning

(CIF, 2013). The Adaptation Fund has allocated US$231.5 million to concrete adaptation projects and programmes in developing countries (Adaptation Fund, 2014).

At the COP-16 conference in Cancun 2010, Annex-I countries committed to disbursing US$30 billion of "fast-start finance" between 2010 and 2012; and to mobilize US$100 billion per annum by 2020 under the Green Climate Fund (GCF), to support both adaptation and mitigation (UNFCCC, 2011). At the COP-18 conference in Doha 2012 and COP-19 in Warsaw 2013; Annex-I countries were encouraged and urged respectively to continue mobilizing finance at increasing levels from fast-start finance levels (UNFCCC, 2013, 2014). This represents a significant pool of finance that can be used to finance adaptation in developing countries.

While this represents a substantial amount of available finance, it has been shown to be inadequate by recent estimations of adaptation finance requirements (cf. Parry et al., 2009; Stern, 2007; UNFCCC, 2007; World Bank, 2006, 2010). These estimations may also underestimate total requirements. The World Bank (2010) estimated the costs of climate-proofing existing infrastructure at US$70–100 billion per year by 2050 with a 2°C temperature rise – a level of warming which could well be exceeded with current mitigation pledges (Hare et al., 2013). If available finance is less than what is needed, this makes the efficient use of resources paramount.

4. Possibilities for financing CBA under existing UNFCCC architecture

CBA is currently financed from a mixture of private sources and public sources such as overseas development assistance (ODA). However, this paper restricts its focus to the multilateral funds used to channel adaptation finance committed under the UNFCCC process. These funds are the LDCF, SCCF, Adaptation Fund (as funds created under the Convention and its Kyoto Protocol), and PPCR. In addition, the Small Grants Programme of the GEF (GEF-SGP) is considered despite not having a specific mandate nor any direct financial donations from countries as part of UNFCCC-related commitments. The GEF-SGP is considered because of its experience with CBA, and because it is funded via the GEF to which funds can be made as part of official climate finance commitments.

One way to assess the availability of finance for CBA under the UNFCCC architecture is by considering the mandates and access modalities of the dominant multilateral actors charged with delivery of adaptation finance.

Finance from the LDCF and SCCF is accessed by developing-country governments through submission of proposals to the GEF via permitted implementing agencies. Funds are allocated to projects responding to priorities identified in NAPAs or that conform to criteria based on guidance from the Convention. The access modalities of

the LDCF and SCCF have been previously criticized (see Ayers & Huq, 2009), with some developing countries eager to propose ways for finance to be channelled more directly (see Talvela & Uitto, 2009).

Regarding the specific role of CBA, the GEF (2006, p. 6) acknowledges that developing countries have identified CBA as a "cross-sectoral priority requiring urgent attention". Review criteria for funding proposals are sufficiently flexible to allow the financing of community-level adaptation activities, but they do not make the financing of such activities obligatory or require the directing of finance to local-level actors. There are cases where LDCF-funded projects and programmes strongly incorporate CBA elements, such as the "Community-based Adaptation to Climate Change through Coastal Afforestation" project carried out in coastal Bangladesh (MoEF, 2008; Rawlani & Sovacool, 2011).

The entity that has thus far raised the largest amount of adaptation finance is the PPCR (Climate Funds Update, 2013) that distributes grants and loans to a selected group of 18 countries considered to be highly vulnerable (CIF, 2009). Finance is accessed through multilateral development banks, which implement programmatic activities in coordination with the national government and other recipient-country stakeholders. Its investment priorities emphasize action at the national level with the aim of integrating climate risk into national and sectoral development plans (CIF, 2009). The focus, therefore, is not on community-level adaptation activities, though recently some country programmes such as in Zambia have been integrating CBA activities (CIF, 2014).

The Adaptation Fund was given a specific mandate under decision 1/CMP.4 to finance concrete adaptation activities with a strategic focus on giving special attention to the needs of the most vulnerable communities (UNFCCC, 2009). Such a mandate may make the Adaptation Fund more closely aligned with the needs of CBA. Like the LDCF and SCCF, the Adaptation Fund allows countries to access funds through multilateral intermediaries who supervise the implementation of activities. But it has reserved half of available funds for "direct access" by National Implementing Entities (NIEs) that become eligible to make applications after passing an accreditation process. To date, over US$44 million has been accessed by NIEs in 5 countries; 12 countries with accredited NIEs have not yet accessed grant finance for programmes through direct access, however, 10 countries are in the process of formulating proposals.

The direct access modality was designed to increase national ownership of adaptation projects and programmes (Brown, Bird, & Schalatek, 2010) and has been praised by civil society as an innovative approach seeking to ensure country ownership, as well as increasing country oversight and accountability (Brown et al., 2010; Horstmann & Abeysinghe, 2011). The process of direct access bypasses multilateral intermediaries and devolves fiduciary risk management powers to national-level entities, either state or non-state actors depending on country preference. Through devolution of management powers, direct access shifts decision-making to national and sub-national levels (Marston, 2013; Mueller, 2013). Direct access, as used by the Adaptation Fund, may therefore represent an opportunity for communities to achieve a greater voice in the allocation of finance to address community-level adaptation needs. An example of an Adaptation Fund financed programme that has helped to up-scale finance for CBA in Senegal is shown in Box 2.

The GEF-SGP has itself distributed over US$14.7 million in grants for CBA since 1992, and has mobilized over US$5 million in total cash co-financing and over US$ 9 million of in-kind co-financing for CBA over this period (GEF-SGP, 2014). The funds for CBA within the GEF-SGP make up a small proportion of the total of US$450 million in grants distributed since 1992, which includes activities such as land degradation and sustainable forest management, however, there are reasonable grounds to assume these activities have elements which can be considered as CBA.

The GEF-SGP supports non-government organization (NGO) and community-based organization (CBO) projects in developing countries on various thematic areas, one of which is adaptation. The CBA component of the GEF-SGP makes grants of up to US$50,000 directly to NGOs or CBOs with the aim of developing community-based capacity and tools to respond to adverse impacts of climate change; supporting CBA projects in selected countries; and disseminating lessons learned at the community level (GEF, 2008).

GEF-SGP activities are implemented through a National Steering Committee and guided by Technical Advisory Groups, comprising civil society, government ministries, GEF country operational focal point, and UN Convention desk officers. Working with a sub-regional modality, and National Steering Committee or National Focus Group, enables GEF-SGP to reach remote communities. The difficulties of working with community groups with little or no capacity are reduced through an initial grant of US$5000 for planning CBA projects. This reduces the risks posed by immediately disbursing relatively large sums while improving the ability of communities to convene, discuss, and plan remedial actions and measures which can provide community-owned and implemented solutions. Risk management techniques during the project execution phase further allow GEF-SGP to retain standards and flexibility compared to other funds which use more technical disbursement processes.

While limited provision exists for community influence within the LDCF, SCCF, AF, and PPCR; it remains unclear what proportion of adaptation finance from these entities is being directed to activities designed or implemented by communities at local levels. However, it seems that the proportion of adaptation finance going through these funds towards CBA is low. Currently, there are few ways for climate-vulnerable stakeholders at the local level to access adaptation finance. Of the funds described above,

only the GEF-SGP has an application mechanism for community groups. In contrast to the direct access window of the Adaptation Fund, finance is channelled by the LDCF, SCCF, and PPCR to intermediary entities, which in turn flows to the national government, to sub-national government, and at times NGO actors. In some cases, NGO actors executing adaptation activities may include community organizations. However, within this way of working, bottlenecks exist in channelling adaptation finance from national to local levels, and only a small proportion may reach local communities (Christensen et al., 2012).

Such a system provides donors with the expectation of strong fiduciary risk management, low transaction costs, and transparency in the delivery of adaptation finance. However, it could be detrimental to ensuring effective solutions for the most vulnerable by narrowing opportunities for communities to access funds to implement local-level adaptation solutions.

5. Increasing financial flows to the local level

The absence of an international commitment to channel a significant proportion of adaptation finance to the local level contradicts with the common understanding that adaptation occurs at this level, and that institutions at this level are important determinants of successful adaptation (Agrawal, 2008; Black, 2010). Without such a commitment, there exists the risk that the significant climate funds being mobilized under the UNFCCC may be used to finance top-down, centralized project activities that may suit the interests of more powerful stakeholders, but struggle to bring benefits to the most vulnerable communities.

At a time when the climate change negotiations centre on achieving a new global agreement by 2015 and financial commitments are to be scaled up to US$100 billion by 2020, there is the expectation from developing countries that climate finance achieve new scales (Griesshaber, 2012). The present time, therefore, is opportune for committing to ensuring that funds reach communities and allow them to implement urgently needed adaptation activities in their local context.

It has been suggested that providing finance to local government institutions, like the case of the potentially replicable Local Adaptation Plans for Action (LAPA) in Nepal (Box 1), could play a role in empowering communities towards CBA (Christensen et al., 2012). The Government of Nepal has recognized the need to ensure that finance reaches the local level, by invoking national-level policies providing that at least 80% of total funds available for climate change activities flow to the local level (GoN, 2011). No such commitment has yet been made internationally at the level of the UNFCCC, nor by funds created under the Convention and its Kyoto Protocol, nor has the international community made moves in this direction through bilateral development agencies. A commitment similar to the Nepalese policy to ensure finance flows to the local level could be instrumental in ensuring that adaptation finance under the UNFCCC reaches vulnerable communities.

Box 1. Ensuring country architecture mainstreams CBA: opportunities and challenges from the LAPA in Nepal.

As one of the last countries to develop its NAPA, Nepal has been able to learn lessons from other countries (Regmi & Karki, 2010). The NAPA process in Nepal emerged from a growing realization that to achieve mainstreaming of adaptation into development across scales, there was a need to link between national-level climate change activities and local planning (Regmi & Karki, 2010).

In November 2011, the Government of Nepal endorsed the LAPA framework, which provides opportunities to implement NAPA priorities with participation of local communities in a bottom-up and inclusive process. Nepal's Climate Change Policy and NAPA documents have made mandatory provisions which intend to disburse at least 80% of the available financial resources directly for local-level implementation of adaptation actions (GoN, 2011).

The LAPA intends to facilitate mainstreaming through the integration of climate change resilience into local-to-national development planning processes, including integration of adaptation priorities into village, municipality, and district planning processes in accordance with the Local Self Governance Act (GoN, 2011). The LAPA process provides opportunities to assess site-specific climate vulnerabilities, identify adaptation options, and implement adaptation actions with participation of local communities. The guiding principles of the LAPA manual are processes that are "bottom-up, inclusive, responsive, and flexible", with Village Development Committees (VDCs) and Municipalities being identified as the most appropriate operational units for LAPA planning (GoN, 2011).

Challenges exist in mainstreaming adaptation finance at the national level in Nepal, including concerns about lack of integration between the NAPA and PPCR finance (Ayers, Kaur, & Anderson, 2011). There can be trade-offs between rapid disbursement and capacity-building, as shown by concerns that the Strategic Priority for Climate Resilience's use of consultants as opposed to Ministry of Environment (MoE) staff achieved a fast turnaround at the expense of government ownership and capacity-building (Ayers et al., 2011).

In addition, there is arguably a trade-off between capacity and ownership. In Nepal, concerns regarding the public financial management system have led to finance being provided through bilateral projects outside the central budget (Wiseman & Chhetri, 2011), leading to concern that capacity and governance constraints are being sidestepped rather than tackled. In the LAPA, there has also been a lack of clarity about how funds will be administered and who will act as trustee (Oxfam, 2011). Due to the fiduciary risk, some major bilateral agencies said they were not able to channel funds through the MoE. The UK's Department for International Development pushed for implementation through managing agents, given MoE capacity constraints (Wiseman & Chhetri, 2011). Meanwhile, MoE stressed that LAPA funding should go through government, channelled to appropriate line agencies (Wiseman & Chhetri, 2011). This demonstrates tensions between donors and government in LAPA implementation. The United Nations Development Programme (UNDP) has now been designated as the implementing agency (Uprety, 2013).

Decentralization is complicated by the risk of corruption in the VDCs which tend to be controlled by local elites (Watt, 2012). Furthermore, VDC institutions can exclude participation of lower-caste communities such as Dalits (Jones & Boyd, 2011). There have been improvements in financial management at the local level, but bottlenecks remain and weak fiduciary systems are a concern (Wiseman & Chhetri, 2011). While limits to local adaptation and institutional capacity to respond exist, this should not undermine the value of input and participation of communities and their institutions, since communities do have knowledge, skills, and wealth to draw upon that will help in lessening dependency on financing and technology (Regmi & Bhandari, 2013). Furthermore, while capacity is certainly a concern, channelling finance and resources through local institutions can itself lead to strengthened capacities (Christensen et al., 2012).

Although operational issues exist, the case of Nepal is important in demonstrating what can be done at the recipient country level to create an institutional environment to facilitate adaptation finance being mainstreamed into CBA and development processes at the local level. CBA is centred around communities' priorities, needs, and strategies, but needs to be embedded in broader strategies (mainstreamed) for wider impact. With this in mind, the LAPA provides an example of combining of bottom-up and top-down adaptation, and demonstrates how sufficient financial flows can be channelled accordingly to reach the local level.

Without such a commitment, delivering significant climate finance from international sources for adaptation at the community level could be achieved through one of two ways: either working within the constraints of existing multilateral systems, or by seeking alternative, parallel channelling systems. Given the urgency with which climate impacts are increasingly being experienced, and the slow pace of fostering consensus on new international agreements, there are benefits to building on the innovative aspects of the climate finance architecture already in place that show potential in terms of up-scaling CBA. The GEF-SGP and the Adaptation Fund demonstrate potential in this respect.

5.1. *GEF-SGP*

At the fund level, the GEF-SGP has been found to have the most specific focus on financing CBA and to be most accommodating to applications by community-level groups. The GEF-SGP has also demonstrated an ability to lead to the up-scaling of finance for CBA and the practice of CBA by other donors and international funds. Projects are required to demonstrate capacity for horizontal replication, as well as provide evidence-based results (GEF, 2012). This has contributed to the mainstreaming of lessons learned into larger adaptation projects or development processes (GEF, 2008, 2012; GEF-SGP, 2013). For instance, lessons learnt in Namibia from GEF-SGP projects have been integrated into the SCCF project in Namibia (GEF, 2013d). Its inclusive approach has led to the dissemination of CBA ideas among national-level policy-makers. For example in Jamaica, the GEF-SGP National Coordinator has been invited by government to contribute to the ongoing climate change strategy and national policy preparations process. GEF-SGP also has experience in channelling finance committed under the Convention as it was previously the delivery mechanism of the CBA component of the now-closed Strategic Priority "Piloting an Operational Approach to Adaptation" (SPA), which was charged to the GEF in 2001. Additionally, it has experience crowding in finance from bilateral sources and other international funds in the form of co-financing (GEF-SGP, 2013).

Despite the promise of the GEF-SGP, adaptation and CBA only form part of its activities and its activities require a strong focus on training for capacity-building, reducing resources available for concrete activities that address urgent adaptation needs. Other limitations include operating at the project rather than programme scale; while available finance is substantially less than in other climate funds. Although up-scaling of successful projects has occurred, this is not always the case. When up-scaling does not occur, existing institutional and fiscal boundaries can restrict initiatives to small "islands of success" (Huq & Faulkner, 2013). Limited community capacity and enabling policy environments potentially hinder attempts to scale-out as fiduciary standards may not be met. While CBOs can apply to the GEF-SGP, local government institutions cannot apply, which may limit the extent to which CBA is mainstreamed at local-level formal government institutions. There are also obvious transaction costs incurred by dealing with many small projects instead of a few large programmes.

However, the GEF-SGP has a clear role to play in piloting tools and activities, demonstrating that these can fit within national policy and planning, and finally demonstrating that channelling finance directly to community-level interventions is viable through the multilateral system. Furthermore, by requiring projects to demonstrate evidence-based results and potential for replication and expansion, the GEF-SGP builds the foundation for horizontal replication of CBA at the local level, and the mainstreaming of CBA in other projects or programmes of other donors and international funds. Increasing finance to the GEF-SGP is a clear option for increasing finance for CBA; however, the lack of a clear mandate from the UNFCCC may frustrate such attempts. One potential option is to use the GEF-SGP as a vehicle through which components of the programmes from other funds can be operationalized. An example of this is the Adaptation Fund programme in the Cook Islands which disbursed small grants through the GEF-SGP.

5.2. *Adaptation Fund*

At the larger scale, and at the level of funds specifically mandated to deliver climate finance, the Adaptation Fund demonstrates the first case of the devolution of decision-making powers to national and sub-national levels. Within the model of direct access, authority over the formulation, design, and implementation of interventions rests entirely with accredited national entities, dispensing with the influence of multilateral intermediaries. In order to access funds, these entities must demonstrate that the adaptation benefits achieved with the help of the project/programme can be sustained after its end, and should enable replication and expansion after its completion through, for example, policy, governance, and other arrangements as required (Adaptation Fund, 2013a).

Box 2. Ensuring international climate funds mainstream CBA – Adaptation Fund and Senegal.

In 2010, Senegal was the first country to be awarded finance from the Adaptation Fund, for a US$8.6 million programme entitled "Adaptation to Coastal Erosion in Vulnerable Areas". The programme was implemented and supervised through the direct access modality by the Centre de Suivi Ecologique (CSE), a Senegalese entity that had demonstrated compliance with the Fund's fiduciary standards. The project sought to protect coastal areas in three locations (Rufisque, Saly, and Joal) against erosion exacerbated by sea level rise; to construct anti-salt dykes to reduce salinization of agricultural land used to grow rice; to assist local communities to manage fish processing; to raise awareness regarding adaptation; and to develop and implement updated regulations that would mainstream climate change into coastal area management.

Two non-governmental entities, Green Senegal and Dynamique Femmes, led the implementation of over US$1.1 million of programme activities. They undertook stakeholder consultations during the design stage of programme activities, and sought inputs from local communities on potential solutions to adaptation challenges (Dossou et al., 2012). These entities had a history of working with local communities, demonstrating the potential for direct access to devolve decision-making and responsibility to the local level.

Specific interventions were decided upon following inputs from affected populations, including women's groups in Joal, as well as vulnerability studies that contributed to development of Senegal's NAPA (Dossou et al., 2012). Such an approach demonstrates a combination of bottom-up influence from community voices in decision-making coupled with top-down information on climate scenarios and adaptation options.

The Senegalese programme can be considered as best practice in mainstreaming climate change from a dual perspective. Firstly, it strikes an effective balance between concrete measures, such as coastal infrastructural measures that address pressing community needs; and elements of training, capacity enhancement, and policy reform that lay the groundwork for longer term transition towards inclusion of climate change into development planning across community to national scales. Secondly, the groundwork put in place by Senegal to get a national entity accredited through the direct access system resulted in strengthened procurement, transparency, and project monitoring (Adaptation Fund, 2012) as well as establishment of mechanisms for community involvement to ensure that the programme benefits the most vulnerable people (Dossou et al., 2012).

The Senegal programme demonstrates a number of important lessons. Its achievements include the substantial completion of coastal protection infrastructure, the close involvement of civil society and community members, and a tailored communication strategy that comprises a diverse range of media from radio programmes to traditional wrestling sessions that communicate adaptation messages.

Challenges faced include ensuring equitable distribution of programme benefits. In an area of Rufisque, community consultations led to the prioritization of vulnerable areas for protection by coastal infrastructure. CSE ameliorated concerns held by adjacent communities who did not directly benefit from the programme by holding consultations with the parties involved.

The experience of Senegal demonstrates the need to manage expectations among potential stakeholders from an early stage, and to ensure difficult decisions are taken transparently with community members being fully informed of the likely implications of the decisions made. The benefits of this particular intervention include the close involvement of communities in decision-making and the concrete outcomes for the most vulnerable communities. Such a legacy could prove invaluable in delivering CBA solutions in Senegal as climate finance evolves.

The 17 national entities accredited to date under the Adaptation Fund include cooperatives, NGOs, research bodies, national trust funds, government agencies, and ministries. To access funds, they are required to give special attention to the most vulnerable communities by demonstrating how any proposed intervention gives them economic, social, and environmental benefits. Applicants must undertake a comprehensive consultation process with all direct and indirect stakeholders, and reflect the results in the proposal design. Despite early problems (see Bird, Billett, & Colón, 2011; Harmeling & Kaloga, 2011) anecdotal evidence from developing countries using direct access suggests that stronger community-level decision-making exists than in similar interventions using conventional delivery mechanisms (Dossou et al., 2012). To date, however, only five entities have accessed project or programme finance through the direct access modality (Adaptation Fund, 2014) and it remains too early to make definitive judgements about its long-term benefits, while the remaining entities continue their efforts to submit proposals on behalf of their countries. Additionally, capacity constraints restrict the ability of potential implementing entities from gaining accreditation – though capacity-building from other actors can overcome this (Bird et al., 2011). Recognizing the challenges that some countries have had in selecting an appropriate candidate entity to become a NIE, the Adaptation Fund Board approved a Readiness Programme (Adaptation Fund, 2013b). The programme, which has an overarching goal of increasing the number of developing countries benefiting from direct access to adaptation finance, aims to increase the number of accredited NIEs from 16 to at least 23, and the number of projects and programmes fully implemented through direct access from 5 to at least 13.

This initiative on readiness for direct access should yield new experiences and new data for analysis of the potential of direct access to promote integration of CBA in national processes throughout developing countries.

The common and innovative characteristics of the GEF-SGP and the Adaptation Fund's direct access modality relate to the decentralization of decision-making power to national and sub-national levels. To varying degrees, depending on the specifics of the intervention, this approach brings communities closer into the process of conceptualizing, formulating, and ultimately implementing and evaluating adaptation. As communities in developing countries and CBA practitioners seek to put more power into the hands of those experiencing climate impacts at the local level, experiences from these funding modalities can potentially be integrated throughout the international climate finance architecture. Overall by requiring proposals to take the concerns of communities into account, the Adaptation Fund helps to mainstream CBA processes from the fund level downwards.

Looking ahead, the GCF has committed to using simplified access to funding, including direct access as a delivery mechanism (GCF, 2013). Likewise, since its fifth replenishment the GEF has used direct access modalities. However, this is different to its interpretation under the Convention (UNFCCC, 2008) and made operational through the Adaptation Fund, where power is expected to be devolved to the national level. Under the GEF, countries can only directly access limited finance for national portfolio formulation exercises and preparation of Convention reports (GEF, 2013b). While of those under consideration for accreditation as implementing entities, only four were not regional- or international-level entities (GEF, 2013c). As deliberations on raising funds and operationalizing access modalities ensue, the international community could look to the GEF-SGP and Adaptation Fund for innovative approaches to ensure that finance for adaptation reaches the most vulnerable at the community level. While efforts to successfully up-scale finance for CBA will be helped by the replication of modalities such as those used by the GEF-SGP and Adaptation Fund, they may not alone be sufficient. CBA is likely to receive more climate finance if it operates within a national-level institutional environment that commits resources to the local level, an example being Nepal's LAPA. A similar commitment at the UNFCCC level or by funds operating under the Convention and its protocol would facilitate the transition to this institutional environment.

6. Conclusion

With finance under the UNFCCC expected to rise, the time is now favourable for exploring ways in which adaptation finance has been channelled from finance institutions to CBA initiatives to date, and how these achievements can be up-scaled. Currently, it is very difficult to calculate the amount of climate finance being channelled towards CBA as this is not being reported. Self-reporting by climate funds would help ameliorate this issue; however, this is inhibited by the lack of a mandate to spend secretariat resources doing so, the absence of an international commitment to dedicate a certain proportion of adaptation finance to local communities, and definitional issues that frustrate the identification of CBA in implemented projects.

This paper has assessed adaptation finance mandates and access modalities in order to explore ways finance is being channelled towards CBA activities. Of the large funds which have received official financial commitments under the UNFCCC, it is the Adaptation Fund that has thus far implemented rules and procedures to promote institutionalized usage of CBA in adaptation programmes. It achieves this through its mandate to serve the needs of the most vulnerable communities, by demanding that proposals give special attention to vulnerable communities; and via its direct access modality, channelling finance through implementing entities more likely to be accountable to local communities. The LDCF, SCCF, and PPCR have financed programmes, which have involved CBA processes, but these processes are not as institutionalized as they are in the rules and procedures of the Adaptation Fund and the CBA component of the GEF-SGP. In short, they neither promote nor exclude CBA in potential programmes.

The GEF-SGP has been found to be most accommodating to applications by community-level groups, while also demonstrating an ability to lead to the up-scaling of CBA. However, finance available in this modality is significantly smaller than other funds explored in this paper. Yet, the GEF-SGP has a clear role to play in piloting evidence-based tools and activities, demonstrating that CBA can fit within national policy and planning, and finally demonstrating that channelling finance directly to communities is a viable option.

It is quite possible that some aspects of the rules and procedures of the Adaptation Fund and GEF-SGP could be incorporated by other UNFCCC funds. Proposals could be required to give special attention to, and implement activities via institutions downwardly accountable to climate-vulnerable stakeholders. However, up-scaling favourable modalities and procedures at fund-level may not alone suffice. There must be efforts to integrate CBA into national policy and planning with a commitment to dedicate resources to local levels. The Nepalese LAPA framework demonstrates a national-level institutional environment that can facilitate this outcome. The LAPA is an example of the integration of bottom-up and top-down approaches to adaptation, demonstrating how financial flows for adaptation can be channelled to local levels. It achieves this by requiring that at least 80% of total funds available for climate activities flow to the grassroots level; and by

requiring that project implementation is conducted with the participation of local communities in a bottom-up and inclusive process. Such a commitment could be made at the UNFCCC level, or the funds operating under the Convention and its Kyoto Protocol. If climate funds under the UNFCCC process enacted similar measures, more appropriate levels of finance are likely to be channelled towards CBA activities.

Acknowledgements

We thank Professor Jouni Paavola, Dr Stavros Afionis, and Dr Hannah Reid for helpful comments on earlier versions of the manuscript.

Disclaimer

This article is the result of a collaboration between academic researchers and practitioners in the field of climate finance. The findings, interpretations, and conclusions expressed by the authors are their own and not of the institutions with whom they are affiliated.

References

Adaptation Fund. (2012). *Report on the learning mission to Senegal*. Retrieved from https://www.adaptation-fund.org/sites/default/files/AFB.EFC_.10.5.Report_of_the_Learning_Mission_to_Senegal.pdf

Adaptation Fund. (2013a). *Instructions for preparing a request for project or programme funding from the Adaptation Fund*. Retrieved from https://www.adaptation-fund.org/sites/default/files/OPG%20ANNEX%204-2%20Instructions%20(Nov2013).pdf

Adaptation Fund. (2013b). *Options for a programme to support readiness for direct access to climate finance for national and regional implementing entities*. Retrieved from https://www.adaptation-fund.org/sites/default/files/AFB.B.22.6%20Options%20for%20a%20climate%20finance%20readiness%20programme%20for%20NIEs%20and%20RIEs_0.pdf

Adaptation Fund. (2014). *Funded projects*. Retrieved from https://www.adaptation-fund.org/funded_projects

Adger, W.N., Huq, S., Brown, K., Conway, D., & Hulme, M. (2003). Adaptation to climate change in the developing world. *Progress in Development Studies, 3*(3), 179–195.

Agrawal, A. (2008). *The role of local institutions in adapting to climate change*. Retrieved from http://www.icarus.info/wp-content/uploads/2009/11/agrawal-adaptation-institutions-livelihoods.pdf

Ayers, J. (2011). Resolving the adaptation paradox: Exploring the potential for deliberative adaptation policy-making in Bangladesh. *Global Environmental Politics, 11*(1), 62–88.

Ayers, J., & Dodman, D. (2010). Climate change adaptation and development I: The state of the debate. *Progress in Development Studies, 10*, 161–168.

Ayers, J., & Forsyth, T. (2009). Community-based adaptation to climate change: Strengthening resilience through development. *Environment: Science and Policy for Sustainable Development, 51*(4), 22–31.

Ayers, J.M., & Huq, S. (2009). Supporting adaptation to climate change: What role for official development assistance?.

Development Policy Review, 27(6), 675–692. doi: 10.1111/j. 1467-7679.2009.00465.x

Ayers, J., Kaur, N., & Anderson, S. (2011). Negotiating climate resilience in Nepal. IDS Bulletin, 42(3), 70–79.

Bird, N., Billett, S., & Colón, C. (2011). Direct access to climate finance: Experiences and lessons learned. ODI/United Nations Development Programme.

Black, E. (2010). Climate change adaptation: Local solutions for a global problem. 22 GEO. INT'L ENVTL. L. REV, 359–360.

Brown, J., Bird, N., & Schalatek, L. (2010). Direct access to the Adaptation Fund: Realising the potential of National Implementing Entities. Climate Finance Policy Brief No. 3, ODI.

Brown, J., & Kaur, N. (2009). Financing adaptation: Matching form with function. Background Note. London: Overseas Development Institute.

Burton, I. (2004). Climate change and the adaptation deficit. In A. Fenech, D. MacIver, H. Auld, R. Bing Rong, & Y. Yin (Eds.), Climate change: Building the adaptive capacity (pp. 25–33). Toronto: Meteorological Service of Canada, Environment Canada.

Christensen, K., Raihan, S., Ahsan, R., Uddin, A.M.N., Ahmed, C.S., & Wright, H. (2012). Financing local adaptation: Ensuring access for the climate vulnerable in Bangladesh. Dhaka: ActionAid Bangladesh, Action Research for Community Adaptation in Bangladesh, Bangladesh Centre for Advanced Studies, and International Centre for Climate Change and Development.

CIF. (2009). The selection of countries to participate in the Pilot Program for Climate Resilience (PPCR) Report of the Expert Group to the Subcommittee of the PPCR. Retrieved July 24, 2013, from https://www.climateinvestmentfunds.org/cif/sites/climateinvestmentfunds.org/files/PPCR_Selection_of_Countries_to_Participate_Report_of_the_Expert_Group_fin al.pdf

CIF. (2013). Pilot program for climate resilience. Retrieved July 25, 2013, from https://www.climateinvestmentfunds.org/cif/Pilot_Program_for_Climate_Resilience

CIF. (2014). Zambia's PPCR strategic program. Retrieved January 15, 2014, from https://www.climateinvestmentfunds.org/cifnet/investment-plan/zambias-ppcr-strategic-program

Climate Funds Update. (2013). Data. Retrieved August 1, 2013, from http://www.climatefundsupdate.org/data

Dossou, K., Ferrera, I., Rodriguez, E., Harmeling, S., Kaloga, A. O., Kakakhel, K.S., … , Seck, E. (2012). Independent insights from vulnerable developing countries. Adaptation Fund NGO Network. Retrieved July 28, 2013, from http://germanwatch.org/en/download/7568.pdf

Ensor, J., & Berger, R. (2010). Governance for community-based adaptation. Rugby: Practical Action, the Schumacher Centre for Technology and Development.

Forsyth, T. (2013). Community-based adaptation: A review of past and future challenges. Wiley Interdisciplinary Reviews: Climate Change, 4(5), 439–446.

GCF. (2013). Mandate and governance. Retrieved July 28, 2013, from http://gcfund.net/about-the-fund/mandate-and-governance.html

GEF. (2006). Programming paper for funding the implementation of NAPAs under the LDC Trust Fund. GEF/C.28/18. Global Environmental Facility.

GEF. (2008). Joint evaluation of the GEF small grants programme (Evaluation Report No. 39). Washington, DC: Global Environment Facility Evaluation Office.

GEF. (2012). Replication and up-scaling. Retrieved January 15, 2013, from https://sgp.undp.org/index.php?option=com_content&view=article&id=117&Itemid=192#.UtZrTPRSZVI

GEF. (2013a). Data mapping portal. Retrieved January 15, 2013, from http://www.thegef.org/gef/RBM

GEF. (2013b). Direct access for national communications and biennial update reports to UNFCCC. Retrieved January 20, 2013, from http://www.thegef.org/gef/CC_direct_access

GEF. (2013c). Progress report on the pilot accreditation of GEF project agencies. Washington, DC: Global Environmental Facility.

GEF. (2013d). Project identification form. Retrieved May 14, 2014, from http://www.thegef.org/gef/sites/thegef.org/files/gef_prj_docs/GEFProjectDocuments/Climate%20Change/Namibia%20-%20(5343)%20-%20Scaling%20Up%20Community%20Resilience%20to%20Climate%20Variabi/ID5343%20%20Resubmission_Namibia%204711%20PIF%20SCCF_9April2013.pdf

GEF-SGP. (2013). SGP is implementing a portfolio of over US $16 million in community based adaptation. Retrieved May 5, 2014, from https://sgp.undp.org/index.php?option=com_content&view=article&id=319&catid=36&Itemid=186#.U3CHwfmSxbc

GEF-SGP. (2014). Community-based adaptation. Retrieved May 1, 2014, from http://sgp.undp.org/index.php?option=com_areaofwork&view=summary&Itemid=177

GoN, Government of Nepal. (2011). National framework on local adaptation plans for action. Government of Nepal, Ministry of Science Technology and Environment, Singha Durbar.

Griesshaber, L. (2012). Green climate fund: Timely action needs early pledges. Retrieved July 28, 2013, from http://germanwatch.org/en/download/6889.pdf

Hare, B., Rocha, M., Jeffery, L., Gütschow, J., Rogelj, J., Schaeffer, M., … , Höhne, N. (2013). Warsaw unpacked: A race to the bottom. Climate Action Tracker. Retrieved from http://climateactiontracker.org/assets/publications/briefing_papers/CAT_Policy_brief_Race_to_the_bottom.pdf

Harmeling, S., & Kaloga, A. (2011). Understanding the political economy of the adaptation fund. IDS Bulletin, 42(3), 23–32.

Horstmann, B., & Abeysinghe, A.C. (2011). The adaptation fund of the Kyoto protocol: A model for financing adaptation to climate change? Climate Law, 2(3), 415–437.

Huq, S., & Faulkner, L. (2013). Taking effective community-based adaptation to scale: An assessment of the GEF small grants programme community-based adaptation project in Namibia. International Centre for Climate Change and Development.

Huq, S., & Reid, H. (2007). Community-based adaptation: A vital approach to the threat climate change poses to the poor. IIED Briefing. London: IIED.

IPCC. (2012). Managing the risks of extreme events and disasters to advance climate change adaptation. A Special Report of Working Groups I and II of the Intergovernmental Panel on Climate Change. Cambridge: Cambridge University Press.

Jones, L., & Boyd, E. (2011). Exploring social barriers to adaptation: Insights from Western Nepal. Global Environmental Change, 21(4), 1262–1274.

Marston, A. (2013). Reaching local actors in climate finance. Exploring local access to the Green Climate Fund. Retrieved July 15, 2013, from http://www.bothends.org/uploaded_files/document/130307_Reaching_local_actors_in_climate_finance_Bot.pdf

MoEF. (Ministry of Environment and Forest, Government of the People's Republic of Bangladesh). (2008). NAPA project document: Community-based adaptation to climate change through coastal afforestation in Bangladesh (Project ID: PIMS 3873). Dhaka: MOEF/UNDP.

Moench, M., & Dixit, A. (2004). Adaptive capacity and livelihood resilience: Adaptive strategies for responding to floods *and droughts in South Asia.* Institute for Social and Environmental Transition, International, Boulder and Institute for Social and Environmental Transition.

Moser, C., & Satterthwaite, D. (2008). *Towards pro-poor adaptation to climate change in the urban centres of low- and middle-income countries.* Global Urban Research Centre Working Paper No. 1.

Mueller, B. (2013). *'Enhanced (Direct) Access' Through '(National) Funding Entities' – Etymology and Examples Information Note on the Green Climate Fund Business Model Framework.* Oxford Institute for Energy Studies. Retrieved July 20, 2013, from http://www.oxfordclimatepolicy.org/publications/documents/EnhancedDirectAccess-04-2013.pdf

Oxfam. (2011). *Owning adaptation: Factsheet Nepal.* Oxfam International. Retrieved from www.oxfam.org

Parry, M., Arnell, N., Berry, P., Dodman, D., Fankhauser, S., Hope, C., ..., Wheeler, T. (2009). *Assessing the costs of adaptation to climate change. A review of the UNFCCC and other recent estimates.* London: International Institute for Environment and Development/London: The Grantham Institute for Climate Change, Imperial College London.

Pelling, M. (2011). *Adaptation to climate change: From resilience to transformation.* London: Routledge.

Rawlani, A., & Sovacool, B.K. (2011). Building responsiveness to climate change through community-based adaptation in Bangladesh. *Mitigation and Adaptation Strategies for Global Change, 16*(8), 845–863.

Regmi, B., & Bhandari, D. (2013). Climate change adaptation in Nepal: Exploring ways to overcome the barriers. *Journal of Forest and Livelihood, 11*(1), 43–61.

Regmi, B., & Karki, G. (2010, July). Local adaptation plans in Nepal. *Tiempo, 76.*

Reid, H., Alam, M., Berger, R., Cannon, T., Huq, S., & Milligan, A. (2009). Community-based adaptation to climate change: An overview. In H. Ashley & A. Milligan (Eds.), *Participatory learning 60: Community-based adaptation to climate change* (Vol. 60, pp. 11–33). International Institute for Environment and Development.

Sabates-Wheeler, R., Mitchell, T., & Ellis, F. (2008). Avoiding repetition: Time for CBA to engage with the livelihoods literature? *IDS Bulletin, 39*(4), 53–59.

Schipper, E., Ayers, J., Reid, H., Huq, S., & Rahman, A. (2014). *Community based adaptation to climate change: Scaling it up.* London: Routledge.

Schultz, K.H. (2012). Financing climate adaptation with a credit mechanism: initial considerations. *Climate Policy, 12,* 187–197.

Smit, B., Pilifosova, O., Burton, I., Challenger, B., Huq, S., Klein, R.J.T., ... , Leary, N.A.. (2001). *Adaption to climate change in the context of sustainable development and equity.* In Climate change 2001 impacts adaptation and vulnerability contribution of working group ii to the third assessment report of the intergovernmental panel on climate change. Geneva: IPCC.

Stern, N.H. (2007). *The economics of climate change: The Stern review.* Cambridge: Cambridge University Press.

Talvela, K., & Uitto, J. (2009). *Evaluation of UNDP work with the least developed countries fund and special climate change fund.* UNDP. Retrieved from http://web.undp.org/evaluation/documents/thematic/ldcf/LDCF-SCCF_Evaluation.pdf

UNEP. (2012). *The emissions gap report 2012.* Nairobi: United Nations Environment Programme (UNEP).

UNFCCC. (2002). *Report of the conference of the parties on its seventh session.* Addendum. FCCC/CP/2001/13/Add.1. Bonn: UNFCCC.

UNFCCC. (2007). *Investment and financial flows to address climate change.*

UNFCCC. (2008). *Report of the Conference of the Parties serving as the meeting of the Parties to the Kyoto Protocol on its third session.* Addendum. Part Two: FCCC/KP/CMP/2007/9/Add.1. Bonn: UNFCCC.

UNFCCC. (2009). *Report of the conference of the parties serving as the meeting of the parties to the Kyoto Protocol on its fourth session.* Addendum. Decision 1/CMP.4, Adaptation Fund, FCCC/KP/CMP/2008/11/Add.2. Bonn: UNFCCC.

UNFCCC. (2011). *Report of the conference of the parties on its sixteenth session.* Addendum. Part Two. FCCC/CP/2010/7/Add.1. Bonn: UNFCCC.

UNFCCC. (2013). *Report of the conference of the parties on its eighteenth session.* Addendum. Part Two. FCCC/CP/2012/8/Add.1. Bonn: UNFCCC.

UNFCCC. (2014). *Report of the conference of the parties on its nineteenth session.* Addendum. FCCC/CP/2013/10/Add.1. Bonn: UNFCCC.

Uprety. (2013). Personal communication with Batu Krishna Uprety, Ministry of Environment, Government of Nepal, Bonn, Germany, June 2013.

Urwin, K., & Jordan, A. (2008). Does public policy support or undermine climate change adaptation? Exploring policy interplay across different scales of governance. *Global Environmental Change, 18,* 180–191.

Van Aalst, M.K., Cannon, T., & Burton, I. (2008). Community level adaptation to climate change: The potential role of participatory community risk assessment. *Global Environmental Change, 18,* 165–179.

Vernon, T. (2008). The economic case for pro-poor adaptation: What do we know? *IDS Bulletin, 39*(4), 32–41.

Watt, R. (2012). Linking national and local adaptation planning: Lessons from Nepal, Case study 3, The Learning Hub, IDS, UK.

Wiseman, K., & Chhetri, R. (2011). *Governance of climate change adaptation finance: Nepal.* Oxfam Research Report.

World Bank. (2006). *Environmentally & socially sustainable development and infrastructure vice presidencies, 'clean energy and development: Towards an investment framework'.* Washington, DC: Development Committee.

World Bank. (2010). *Economics of adaptation to climate change: Synthesis report.* Washington, DC: The World Bank.

Yamin, F., Rahman, A., & Huq, S. (2005). Vulnerability, adaptation and climate disasters: A conceptual overview. *IDS Bulletin, 36*(4), 1–14.

Index

For Product Safety Concerns and Information please contact our EU
representative GPSR@taylorandfrancis.com
Taylor & Francis Verlag GmbH, Kaufingerstraße 24, 80331 München, Germany

www.ingramcontent.com/pod-product-compliance
Lightning Source LLC
Chambersburg PA
CBHW080243270326
41926CB00020B/4350

9 781138 294936